Methods in
Food Analysis

Methods in Food Analysis

Editors

Rui M.S. Cruz
CIQA and Department of Food Engineering
ISE, University of Algarve, Portugal

Igor Khmelinskii
CIQA and Department of Chemistry and Pharmacy
FCT, University of Algarve, Portugal

Margarida C. Vieira
CIQA and Department of Food Engineering
ISE, University of Algarve, Portugal

CRC Press is an imprint of the
Taylor & Francis Group, an **informa** business

A SCIENCE PUBLISHERS BOOK

CRC Press
Taylor & Francis Group
6000 Broken Sound Parkway NW, Suite 300
Boca Raton, FL 33487-2742

© 2014 Copyright reserved
CRC Press is an imprint of Taylor & Francis Group, an Informa business

No claim to original U.S. Government works

International Standard Book Number: 978-1-4822-3195-3 (Hardback)

This book contains information obtained from authentic and highly regarded sources. Reasonable efforts have been made to publish reliable data and information, but the author and publisher cannot assume responsibility for the validity of all materials or the consequences of their use. The authors and publishers have attempted to trace the copyright holders of all material reproduced in this publication and apologize to copyright holders if permission to publish in this form has not been obtained. If any copyright material has not been acknowledged please write and let us know so we may rectify in any future reprint.

Except as permitted under U.S. Copyright Law, no part of this book may be reprinted, reproduced, transmitted, or utilized in any form by any electronic, mechanical, or other means, now known or hereafter invented, including photocopying, microfilming, and recording, or in any information storage or retrieval system, without written permission from the publishers.

For permission to photocopy or use material electronically from this work, please access www.copyright.com (http://www.copyright.com/) or contact the Copyright Clearance Center, Inc. (CCC), 222 Rosewood Drive, Danvers, MA 01923, 978-750-8400. CCC is a not-for-profit organization that provides licenses and registration for a variety of users. For organizations that have been granted a photocopy license by the CCC, a separate system of payment has been arranged.

Trademark Notice: Product or corporate names may be trademarks or registered trademarks, and are used only for identification and explanation without intent to infringe.

Library of Congress Cataloging-in-Publication Data

Methods in food analysis / editors, Rui M.S. Cruz, Igor Khmelinskii, Margarida C. Vieira.
       p. ; cm.
   Includes bibliographical references and index.
   ISBN 978-1-4822-3195-3 (hardcover : alk. paper)
   I. Cruz, Rui M. S., editor of compilation. II. Khmelinskii, Igor, editor of compilation. III. Vieira, Maria Margarida Cortez, editor of compilation.
   [DNLM: 1. Food Analysis--methods. QU 50]
   RA601
   363.19'26--dc23
                                                        2013047504

Visit the Taylor & Francis Web site at
http://www.taylorandfrancis.com

CRC Press Web site at
http://www.crcpress.com

Science Publishers Web site at
http://www.scipub.net

Preface

Measurements of food quality parameters, such as physical, chemical, microbiological and sensory parameters are necessary to characterize both existing and newly developed food products, to avoid possible adulterations/contaminations, and thus, control their quality at every stage of production/distribution or storage at industrial and laboratory scales. Several methodologies are reported in literature that allow quantifying different quality parameters. This book comprehensively reviews methods of analysis and detection in the area of food science and technology. It covers topics such as lipids, color, texture and rheological properties in different food products. The book focuses on the most common methods of analysis, presenting methodologies with specific work conditions. The book is divided into seven chapters, each dealing with the determination/quantification analyses of quality parameters in food products.

    It is an ideal reference source for university students, food engineers and researchers from R&D laboratories working in the area of food science and technology. This book is also recommended for students at undergraduate and postgraduate levels in food science and technology.

    The editors would like to express their sincere gratitude to all contributors of this book, for their effort to complete this valuable venture.

Rui M.S. Cruz
Igor Khmelinskii
Margarida C. Vieira

Contents

Preface   v

1. **Textural and Rheological Properties of Fruit and Vegetables**   1
   R.K. Vishwakarma, Rupesh S. Chavan, U.S. Shivhare and *Santanu Basu*
   1.1 Introduction   2
   1.2 Concepts of Stress and Strain   3
   1.3 Rheology   4
       1.3.1 Shear Stress   4
       1.3.2 Shear Strain   4
       1.3.3 Shear Rate   4
       1.3.4 Viscosity and Apparent Viscosity   5
       1.3.5 Shear Modulus   7
   1.4 Texture of Solids   7
       1.4.1 Stress-Strain Relationship   8
       1.4.2 Compression Test of Food Materials   9
       1.4.3 Stress Relaxation   14
       1.4.4 Creep   15
       1.4.5 Deformation Testing Using Other Geometries   15
       1.4.6 Tensile Loading   15
       1.4.7 Fracture Test   17
       1.4.8 Cutting and Shearing Test   17
       1.4.9 Bending and Snapping Test   17
       1.4.10 Puncture and Penetration Test   19
       1.4.11 Texture Profile Analysis (TPA)   19
       1.4.12 Torsional Loading   21
       1.4.13 Test Specimen and Testing Conditions   21
   1.5 Steady State Rheology   22
       1.5.1 Time Dependent Rheology   23
   1.6 Viscoelasticity   25
       1.6.1 Dynamic Rheology   26
       1.6.2 Analysis of Dynamic Rheological Data   27
       1.6.3 Gel Strength and Relaxation Exponent   30
   1.7 Rheometery   30
       1.7.1 Cone and Plate Viscometers   30

1.7.2 Plate and Plate Viscometers	31
1.7.3 Concentric Cylinders	32
1.8 Rheology of Fruit and Vegetable Products	32
1.8.1 Fruit Juices	32
1.8.2 Jams	34
1.8.3 Puree	35
1.8.4 Paste	37
1.8.5 Pulps	38
1.9 Rheology, Texture and Product Quality	39
1.10 Conclusion	40
References	40

2. Pigments and Color of Muscle Foods — 44
Jin-Yeon Jeong, Gap-Don Kim, Han-Sul Yang and Seon-Tea Joo

2.1 Introduction	45
2.2 Pigments Concentration in Muscles	45
2.3 Myoglobin Chemistry	48
2.3.1 Myoglobin and Derivatives	48
2.3.2 Metmyoglobin Reduction	51
2.4 Measurement of Pigments and Meat Color	52
2.4.1 Reflectance Measurements	53
2.4.2 Visual Evaluation	54
2.4.3 Instrumental Color Measurement	54
2.4.4 Computer Vision Analysis	56
2.5 Conclusion	57
References	58

3. Methodologies to Analyze and Quantify Lipids in Fruit and Vegetable Matrices — 62
Hajer Trabelsi and Sadok Boukhchina

3.1 Introduction	62
3.2 Methods for Vegetable Oil Extraction	63
3.3 Thin-layer Chromatography in Lipid Analysis	64
3.4 Gas Chromatography in Lipid Analysis	65
3.5 High Performance Liquid Chromatography (HPLC) in Lipid Analysis	68
3.6 Mass Spectrometric Based Methods for Vegetable Oil Analysis	69
3.7 Raman Spectroscopy for Vegetable Lipid Analysis	71
3.8 Nuclear Magnetic Resonance (NMR)	71
3.9 Capillary Electrophoresis	72
3.10 Conclusion	72
References	72

4. Texture in Meat and Fish Products	**76**
Purificación García-Segovia, Mª Jesús Pagán Moreno	
and *Javier Martínez-Monzó*	
4.1 Introduction	77
4.2 Measuring Texture: The Basis of Test Methods	77
4.3 Guidelines for Measuring Meat and Fish Texture	78
4.3.1 Shearing Test	80
4.3.2 Compression Test	96
4.3.3 Penetration Test	101
4.3.4 Other Texture Methods	102
4.4 Conclusion	104
References	104
5. Pigments in Fruit and Vegetables	**110**
Sara M. Oliveira, Cristina L.M. Silva and *Teresa R.S. Brandão*	
5.1 Introduction	111
5.2 Pigments Extraction	115
5.2.1 Carotenoids and Chlorophylls	115
5.2.2 Anthocyanins	116
5.2.3 Betalains	118
5.3 Methodologies for Pigments Assessment	118
5.3.1 Chromatographic Methods	119
5.3.2 Non-chromatographic Techniques	126
5.4 Conclusion	130
Acknowledgements	131
References	131
6. Lipids in Meat and Seafood	**142**
Rui Pedrosa, Carla Tecelão and *Maria M. Gil*	
6.1 Introduction	143
6.1.1 Main Roles and Structure of Lipids	143
6.1.2 Lipids in Meat	144
6.1.3 Lipids in Seafood	146
6.1.4 Omega 3 and Health	150
6.2 Lipid Extraction Methods	152
6.2.1 Sample Preparation	154
6.2.2 Liquid-Liquid Extractions	161
6.2.3 Solid-Liquid Extractions	165
6.2.4 Lipid Extraction with Nonorganic Solvents	166
6.3 Analysis of Lipid Extracts from Fish and Meat Samples	168
6.3.1 Classical Analytical Procedures	168
6.3.2 Instrumental Methods for Lipid Characterization	185
6.4 Conclusion	191
References	191

7. Vibrational and Electronic Spectroscopy and Chemometrics in Analysis of Edible Oils — 201

Ewa Sikorska, Igor Khmelinskii and Marek Sikorski

7.1 Introduction	202
7.2 Spectral Characteristics of Edible Oils	202
7.2.1 Overview of Spectroscopic Techniques	202
7.2.2 Vibrational Spectroscopy of Oils	203
7.2.3 Electronic Spectroscopy of Oils	210
7.3 Chemometric Analysis of Spectra of Edible Oils	212
7.3.1 General Ideas	212
7.3.2 Exploratory Analysis (PCA, CA)	213
7.3.3 Multivariate Quantitative and Qualitative Models	214
7.4 Application of Spectroscopy and Chemometrics in the Analysis of Edible Oils	222
7.5 Conclusion	229
Acknowledgements	230
References	230
Index	**235**
Color Plate Section	**237**

1

Textural and Rheological Properties of Fruit and Vegetables

R.K. Vishwakarma,[1] Rupesh S. Chavan,[2] U.S. Shivhare[3] and Santanu Basu[3,*]

ABSTRACT

Texture and rheology properties, such as viscosity, are key properties which consumers evaluate while determining the quality and acceptability of fruit and vegetables and manufactured products. Understanding food texture requires an integration of the physical, physiological and psychophysical elements of oral processing. The knowledge of texture and rheology has many applications in the food industry, i.e., from designing aspects of equipment, bulk handling systems, to new product development and quality control of foods. Texture can be defined as "the group of physical characteristics that arise from the structural elements of the food, and are sensed primarily by the feeling of touch, are related to the deformation, disintegration and flow of the food under a force, and are measured objectively by functions of mass, time, and distance". The rheological behavior of fluid foods is determined by measurements of shear stress versus shear rate/time or elastic/viscous modulus versus frequency,

---

[1] Central Institute of Postharvest Technology, P.O. PAU, Ludhiana, Punjab, India.
[2] National Institute of Food Technology, Entrepreneurship and Management, Plot No. 97, Sector 56, HSIIDC Industrial Estate, Kundli, Haryana, India.
[3] Dr. SS Bhatnagar University Institute of Chemical Engineering & Technology, Panjab University, Chandigarh 160014, India.
* Corresponding author

and representation of the experimental data by viscometric/oscillatory diagrams and empirical equations, as a function of temperature and/or concentration. The textural and rheological properties of fruit and vegetables are extremely important for plant physiologists, horticulturists, food scientists and agricultural/food engineers due to different reasons.

1.1 Introduction

Food quality is adjudged by four principal factors comprising (i) appearance, (ii) flavour, (iii) texture, and (iv) nutrition. The first three are termed as 'sensory acceptability factors' because they are perceived by the senses directly. Nutritional quality is not perceived by the senses. The quality factor 'texture' is defined as all the mechanical (geometrical and surface) attributes of a food product perceptible by means of mechanical, tactile and, where appropriate, visual and auditory receptors (ISO 5492 2008). Texture is also defined as human physiological-psychological perception of a number of rheological and other properties of foods and their interactions (McCarthy 1987). According to Bourne, the *textural properties* of a food are the "group of physical characteristics that arise from the structural elements of the food that are sensed by the feeling of touch, are related to the deformation, disintegration, and flow of the food under a force, and are measured objectively by functions of mass, time, and distance" (Bourne 1982). The terms texture, rheology, consistency, and viscosity are often used interchangeably, despite the fact that they describe properties that are somewhat different. In practice the term texture is used primarily with reference to solid or semi-solid foods rather than liquids.

Rheology is the science of deformation and flow of matter. It is the study of the manner in which a fluid responds to applied stress or strain (Steffe 1996). Rheology may be defined as the study of deformation and flow of matter or the response of materials to stress (Bourne 1993). Rheology of a product is related to the flow of fluids and the deformation of matter. The science of rheology has many applications in design of food processing equipment and handling systems such as pumps, piping, heat exchangers, evaporators, sterilizers, and mixers, as well as in product development and quality control of foods (Saravacos 1970; Rao 1977; 1987). A number of food processing operations depend greatly upon rheological properties of the product at an intermediate stage of manufacture because this has a profound effect upon the quality of the finished product. The microstructure of a product can also be correlated with its rheological behaviour allowing development of newer products. In particular, food rheologists have made unique contributions to the study of mouth-feel and its relation to basic rheological parameters (Rao 1986).

Determination and evaluation of the textural properties of solid foods present many difficulties to scientist, as described by Prins and Bloksma (1983). Agricultural materials are generally inhomogeneous, anisotropic and inelastic in nature. Therefore, their behaviour under loads may vary and stresses may affect other parts of material rather than affecting homogeneous materials such as metals. Since agricultural materials and food products deform in response to applied forces, the nature of the response varies widely among different materials. It depends upon many factors including the rate at which force is applied, the previous history of loading, the moisture content and the composition. The force required to produce a given amount of deformation may be used to quantitatively evaluate the texture of raw and processed foods. In case of raw produce, it may be used for developing new varieties, as a criterion for determining those varieties that have desirable texture. Force-deformation testing is used to study damage which occurs during harvesting, handling, and processing. Such studies often give insight into the specific circumstances that lead to failure and how such failure may be prevented. Sample history can be crucial, with results dependent not only on factors such as rate and extent of deformation but also on the sample history, which includes processing and storage effects prior to measurement.

Knowledge of mechanical properties is useful for both manufacturers and users of food-processing equipment. Knowing the ultimate resistance of a product to mechanical loads helps in saving the products from mechanical damage such as bruising (Brusewitz et al. 1991). On the other hand, food processors need to apply required loads for work to be done. For example, applying the minimum force to cut the peel at the peeling stage of food processing is a matter of importance. Knowing the minimum load helps producers save energy and optimise equipment design. Therefore adequate knowledge of the textural and rheological properties of fruit and vegetable (or their transformation products) is important for storage conditions, process equipment design and quality control.

1.2 Concepts of Stress and Strain

Food materials will deform or flow on application of stress. Stress (σ) is defined as the force (F, N) divided by the area (A, m^2) over which the force is applied, and is generally expressed as Pa. Direction of the force with respect to the surface area impacted determines the type of stress. If the force is directly perpendicular to the surface, a normal stress develops tension or compression in the material. If the force acts in parallel to the sample surface, shear stress is experienced.

Strain is a dimensionless quantity representing the relative deformation of a material. The direction of the applied stress with respect to the material

4 Methods in Food Analysis

surface determines the type of strain. If the stress is normal (perpendicular) to a sample surface, the material will experience normal strain (ε) (Steffe 1996; Daubert and Foegeding 1998). Foods show normal strains when they are compressed (compressive stress) or stretched (tensile stress). Normal strain (ε) may be calculated as a true strain by integration over the deformed length of the material.

$$\varepsilon = \int_{L_i}^{L_i + \Delta_L} \frac{dL}{L} = \ln\left(1 + \frac{\Delta_L}{L_i}\right) \qquad \text{eq. (1.1)}$$

1.3 Rheology

The principles of rheology are commonly applied to understand and improve the flow behavior and textural attributes of food materials and to reveal relationships between the physical properties and the functionality of the material (Steffe 1996). Rheology attempts to build relationships between forces and corresponding deformations, and is expressed more fundamentally as shear stress and shear strain.

1.3.1 Shear Stress

Shear stress (τ) is defined as a force (F, N) per unit area (A, m²). Stress is commonly given in Pascal (N/m²), and expressed as

$$\tau = \frac{F}{A} \qquad \text{eq. (1.2)}$$

Shear stress is applied when the force is tangential to the material surface.

1.3.2 Shear Strain

Shear strain occurs when stress is applied parallel to the material surface. Shear strain (γ) is the inverse tangent of the change in distance (Δd) divided by the initial height (h) of the material.

$$\gamma = \tan^{-1}\left(\frac{\Delta d}{h}\right) \qquad \text{eq. (1.3)}$$

1.3.3 Shear Rate

For fluids, a shear stress can induce a unique type of flow called shear flow. The differential change of strain (γ) with respect to time (t) is known as shear (strain) rate ($\dot{\gamma}$, s⁻¹).

$$\dot{\gamma} = \frac{d\gamma}{dt} \qquad \text{eq. (1.4)}$$

1.3.4 Viscosity and Apparent Viscosity

Viscosity, also called dynamic viscosity or absolute viscosity, of a fluid is essentially its internal friction to flow, and rheology provides information about the internal molecular structure of a system. Viscosity (η, Pa.s) is defined as

$$\eta = \frac{\tau}{\dot{\gamma}} \qquad \text{eq. (1.5)}$$

Rheological behavior of fluids is characterized by measurement of viscosity. If a plot of shear stress (τ) vs shear rate ($\dot{\gamma}$) results in a straight line, viscosity (η) is constant and that material is classified as Newtonian (Fig. 1.1). Fluid that does not obey this relationship ($\tau = \eta.\dot{\gamma}$), is non-Newtonian, which includes most of the food materials. According to a standard classification of non-Newtonian fluids or flow behavior, there are three main classes: time independent (steady state), time dependent, and viscoelastic fluids, where the flow is viscoelastic (Fig. 1.1).

Apparent viscosity (η_a, Pa.s) is the measure of resistance to flow or the fluidity of a non-Newtonian fluid and is the ratio of shear stress to shear rate. It is a coefficient calculated using eq. (1.5) from empirical data as if the fluid obeyed Newton's law. Most materials exhibit a combination of two or more types of non-Newtonian behaviour (Lapasin and Pricl 1995; Steffe 1996).

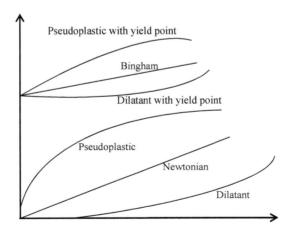

Figure 1.1 Time-independent flow behavior.

Time-independent fluids are materials with flow properties that are independent of the duration of shearing. These fluids are further subdivided into three distinct types:

Shear-thinning or pseudoplastic fluids are characterized by an apparent viscosity which decreases with the increasing shear rate, while the curve begins at the origin (Fig. 1.1). The rate of decrease in viscosity is material-specific.

Shear-thickening fluids, also known as dilatant materials, are characterized by an apparent viscosity that increases with shear rate. This trend is fairly rare in foods.

Viscoplastics fluids are those that exhibit yield stress (τ_0), which is a unique feature of plastic behaviour. Yield stress is a limiting shear stress at which the material begins to flow, while below this yield value the material behaves as an elastic solid.

Time-dependent fluids are materials in which the shear flow properties depend on both the rate and the time of shearing. There are many food products that recover the original apparent viscosity after a sufficient period of rest, while in others the change is irreversible. This type of fluid behaviour may be further divided in two categories: thixotropic and rheopectic.

Thixotropic fluids are characterized by an apparent viscosity that decreases with time when sheared at a constant shear rate. The change in apparent viscosity is reversible, that is, the fluid will revert to its original state on rest. During shearing, the apparent viscosity of the system decreases with time until a constant value is reached and this value typically corresponds to the point where there is no further breakdown of structure. Examples: jam, jelly, marmalade, fruit pulp/juice, cheese etc.

Rheopectic (or anti-thixotropic) fluids are materials in which the apparent viscosity of fluid increases with time when subjected to a constant shear rate. This phenomenon is often an indication of aggregation or gelation that may result from increasing the frequency of collisions or a more favourable position of particles. Examples are rare in food, one is such as starch solution under heating.

Viscoelastic fluids are materials that are simultaneously viscous and elastic. Most food materials exhibit some viscous and some elastic behaviour simultaneously and are therefore referred to as viscoelastic (Gunasekaran and Ak 2000). The viscoelastic properties of materials may be determined using dynamic or transient methods. The dynamic methods include frequency sweep and stress/strain sweep. The transient methods include stress relaxation (application of constant and instantaneous strain and measuring decaying stress with respect to time) and creep (application of constant and instantaneous stress and measuring increasing strain with time).

1.3.5 Shear Modulus

Shear modulus (*G*, Pa) is the constant of proportionality used to relate shear stress with shear strain (Steffe 1996).

$$G = \frac{\tau}{\gamma}$$

eq. (1.6)

1.4 Texture of Solids

Strength, hardness, toughness, elasticity, plasticity, brittleness, ductility and malleability are mechanical properties used as measures of metal behavior under load. However, for all practical purposes, elasticity, hardness, plasticity and brittleness are important mechanical properties for food materials as well. These properties are described in terms of the types of force or stress that the material must withstand and how these are resisted. Common types of stress are compression, tension, shear, torsion, impact or a combination of these stresses, such as fatigue (Fig. 1.2).

Compressive stresses develop within a material when forces compress or crush the material. When a food material is placed between two plates and plates are moved towards each other, the food material is under compression.

Tension (or tensile) stresses develop when a material is subject to a pulling load; for example, using a wire rope to lift a load or when anchoring an antenna. Tensile strength is defined as resistance to longitudinal stress or pull.

Shear stresses occur within a material when external forces are applied along parallel lines in opposite directions. Shear forces can separate material by sliding part of it in one direction and the rest in the opposite direction. When dealing with maximum strength, it is imperative to state the type of loading. A material that is stressed repeatedly usually fails at a point considerably below its maximum strength in tension, compression, or shear.

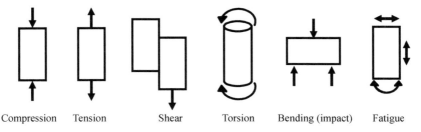

Figure 1.2 Stress applied to a material.

8 Methods in Food Analysis

For example, a noodle can be broken by hand by bending it back and forth several times in the same place; however, if the same force is applied in a steady motion (not bent back and forth), the noodle cannot be broken. The tendency of a material to fail after repeated bending at the same point is known as fatigue.

1.4.1 Stress-Strain Relationship

Rheologically, the question as to whether a particular food is a solid or a liquid is considered in terms of the non-dimensional Deborah number (D), defined by Reiner (1964) as the ratio of the relaxation time of the sample divided by the time of observation. The difference between solids and fluids is then described by the magnitude of D. Time of observation is, in general, a crucial variable when investigating the mechanical properties of foods. If the time of observation is very long or, conversely, if the time of relaxation of the material under observation is very small, the material will flow and will be liquid. On the other hand, if the time of relaxation of the material is larger than time of observation, the material, for all practical purposes, is a solid. It is thus necessary to determine not just a stress-strain curve but the stress-strain-time relationships describing the behaviour of the material. For complex foods there is an artificial distinction between solid and liquid states, which depends not only on the material but also on the experimental time scale relevant to the specific use of the food or the specific process to which the food is subjected.

When force is applied to a solid material and the resulting stress versus strain curve is a straight line through the origin, the material is obeying Hook's law. The relationship may be stated for compressive stress and strain as

$$E = \frac{\sigma}{\delta} \qquad \text{eq. (1.7)}$$

where E is modulus of elasticity (Pa); σ is stress (Pa); and, δ is strain (dimensionless).

The constant E is also known as Young's modulus of elasticity and describes the capability of a material to withstand load. Hookean materials do not flow and are linearly elastic. Strain remains constant until stress is removed and material returns to its original shape. However, most of the food materials follow the Hook's law for small strains only, typically below 0.01. Large strains often produce brittle fracture or non-linear behaviour.

In addition to Young's modulus of elasticity, Poisson's ratio (v) is determined from the compression tests.

$$v = \frac{\text{Lateral strain}}{\text{Axial strain}} \qquad \text{eq. (1.8)}$$

Poisson's ratio may range from 0 to 0.5. Typically v varies from 0.0 for rigid-like materials containing large amounts of air to near 0.5 for liquid-like materials. Values from 0.2 to 0.5 are common for biological materials with 0.5 representing an incompressible substance like potato flesh.

1.4.2 Compression Test of Food Materials

Uniaxial compression is a popular method of testing agricultural materials because the shape of the specimen simplifies the calculation of normal stresses and modulus of elasticity. Since food materials are non-homogeneous, the term apparent modulus of elasticity is used in place of modulus of elasticity. Mohsenin (1986) observed that under small strains, most agricultural materials exhibit extensive elasticity, to which Hertz's theory of contact stress is applicable. The original analysis of elastic contact stresses, by H. Hertz, was published in 1881 and later translated into English by Jones and Schott (Hertz 1896). Deflection occurs when a collinear pair of forces presses the two elliptical bodies together and the point of contact is replaced by a small elliptical area of contact (Hertz 1896) (Fig. 1.3). The equations simplify when the contact area is circular such as

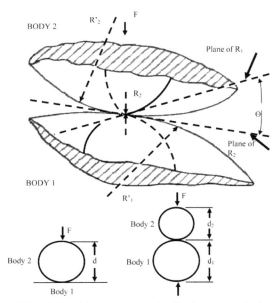

Figure 1.3 Hertz problem (above) for two convex bodies in contact, (below left) sphere on a flat plate, (below right) sphere on sphere (*Adapted from Mohsenin 1986*).

with two spheres or sphere and plate whose principal plane of curvature coincide. To solve the problem, the size and shape of contact area as well as the distribution of normal pressure acting on the area are determined. The deflections and subsurface stresses resulting from the contact pressure are then evaluated with certain fundamental assumptions made to solve the problem (Hertz 1896; Mohsenin 1986). These assumptions are: (i) the material is homogeneous; (ii) contact stress is over a small area relative to the material size; (iii) radii of curvature of the contacting surfaces are substantially greater than radius of the contact area; and (iv) the surfaces are smooth.

Determination of compressive properties requires the production of a complete force-deformation curve. From the force-deformation curve, stiffness, apparent modulus of elasticity, toughness, force and deformation to points of inflection, to bio-yield, and to rupture, work to point of inflection, to bio-yield, and to rupture, and the maximum normal contact stress at low levels of deformation may be obtained. Any number of these mechanical properties can be chosen for the purpose of evaluation and quality control.

When a food material is subjected to compression, it may rupture after following a straight force-deformation curve (Fig. 1.4) (ASAE 1998). The point at which rupture takes place is known as bio-yield point. It is the point where an increase in deformation results in a decrease or no change in force (Fig. 1.4). In a brittle material, rupture may occur in the early portion of the force-deformation curve beyond the linear limit, while it may take place after considerable plastic flow in a tough material (Mohsenin 1986).

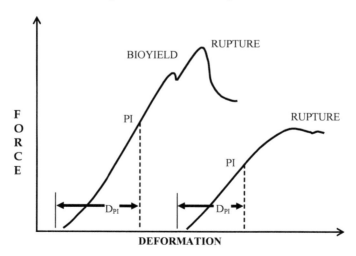

Figure 1.4 Force-deformation curves for materials with and without bioyield point. PI=point of inflection, D_{PI}=deformation at point of inflection (*Adapted from ASAE 1998*).

Toughness is defined as the ability of a material to absorb energy before fracture. This can be approximated by the area under the stress-strain or force-deformation curve up to the point of rupture (Mohsenin 1986).

Apparent Modulus of Elasticity

The apparent modulus of elasticity of the bodies of convex shape may be determined using Hertz equations (Seely and Smith 1965; Timoshenko and Goodier 1970; Mohsenin 1986). It is always determined before the point of inflection. The point of inflection is the point at which rate of change of slope of the curve becomes zero (Fig. 1.4). The combined deformation (δ) of the two bodies along the axis of load is expressed as:

$$\delta = \frac{k}{2}\left\{\frac{9F^2}{\pi^2 E_c^2}\left(\frac{1}{R_1}+\frac{1}{R_1'}+\frac{1}{R_2}+\frac{1}{R_2'}\right)\right\}^{1/3} \qquad \text{eq. (1.9)}$$

where R_1 is the maximum radius of curvature of the body 1 (mm); R_1' is the minimum radius of curvature of body 1 (mm); R_2 is the maximum radius of curvature of body 2 (mm); R_2' is the minimum radius of curvature of body 2 (mm); k is a factor depending on the curvature of bodies (dimensionless); F is force applied (N); δ is combined deformation of both bodies (mm); and, E_c is apparent modulus of elasticity (MPa).

The E_c is the contact modulus and is expressed as:

$$\frac{1}{E_c} = \frac{1-v_1^2}{E_1}+\frac{1-v_2^2}{E_2} \qquad \text{eq. (1.10)}$$

where v_1 and v_2 are Poisson's ratios of bodies 1 and 2; and E_1 and E_2 are apparent moduli of elasticity of body 1 and 2 respectively.

The major and minor axes of the elliptical contact area can be calculated using equations (1.11) and (1.12).

$$\alpha = m\left[\frac{3F}{2E_c}\left(\frac{1}{R_1}+\frac{1}{R_1'}+\frac{1}{R_2}+\frac{1}{R_2'}\right)^{-1}\right]^{1/3} \qquad \text{eq. (1.11)}$$

$$\beta = n\left[\frac{3F}{2E_c}\left(\frac{1}{R_1}+\frac{1}{R_1'}+\frac{1}{R_2}+\frac{1}{R_2'}\right)^{-1}\right]^{1/3} \qquad \text{eq. (1.12)}$$

The values of k, m, and n depend on the principal curvatures of the bodies at the point of contact and the angle between the normal planes containing the principal curvatures. The values of k, m, and n are available in literature for various values of angle (θ) between the normal planes

12 Methods in Food Analysis

containing the curvatures (Timoshenko and Goodier 1970; Mohsenin 1986; ASAE 1998).

The maximum contact stress occurs at the centre of the surface of contact (the first point of contact between the compression tool and the sample). It is numerically equal to 1.5 times the average contact pressure and can be calculated from following equation

$$\sigma_{max} = \frac{1.5F}{\pi\alpha\beta} \qquad \text{eq. (1.13)}$$

In case of nearly spherical food materials (approximated as a sphere compressed between two large rigid plates, with the principal planes of curvature coinciding), the following is valid:

(i) for a spherical body (plant seed) with diameter D_g: $R_1=R_1'=R=D_g/2$;
(ii) for the flat plate: $R_2=R_2'=\infty$; and
(iii) for $\theta = 90°$: $m=n=1; k=1.3514$

For the special case of a rigid plate of metal, E_2 (of compression tool) is much higher than E_1 (of material). The contact modulus can therefore be expressed as:

$$\frac{1}{E_c} = \frac{1-v_1^2}{E_1} \qquad \text{eq. (1.14)}$$

Using these values in eq. (1.7) and rearranging, the apparent modulus of elasticity of the material is expressed as:

$$E_1 = \frac{0.338F(1-v_1^2)k^{3/2}}{\delta^{3/2}}\left[\frac{2}{R}\right]^{1/2} \qquad \text{eq. (1.15)}$$

Bio-yield Point for Spherical Materials

The Hertz equations are used to predict the failure of food materials under quasi-static compressive loading. At bio-yield point, radius of contact circle (α) is computed using equation (1.16) (Timoshenko and Goodier 1970; Shigley and Mishke 2001).

$$\alpha = \left(\frac{3FR_1(1-v_1^2)}{4E_1}\right)^{1/3} \qquad \text{eq. (1.16)}$$

The maximum stress occurs on the axis of loading at the centre of the contact area where the two bodies first come into contact. It is numerically equal to 1.5 times the mean stress and is given by equation (1.17) (Seely and Smith 1965; Timoshenko and Goodier 1970; Shigley and Mishke 2001).

$$\sigma_{max} = \left(\frac{3F}{2\pi\alpha^2}\right)$$

eq. (1.17)

The material tends to expand in the *x*- and *y*- directions when compressed normal to the axis of compression (*z*-direction). The surrounding material, however, does not permit the expansion, and compressive stresses are produced in *x*- and *y*-directions. The two planes of symmetry in loading and the spherical geometry dictate that principal stresses $\sigma_x = \sigma_y$ and $\sigma_z = \sigma_{max}$ occur at the point of contact. The principal stresses at a distance *z* below the surface along the compression axis are given by the following expressions (Timoshenko and Goodier 1970; Shigley and Mishke 2001).

$$\sigma_x = \sigma_y = -\sigma_{max}\left[\left(1-\frac{z}{\alpha}\tan^{-1}\frac{1}{z/\alpha}\right)(1+v_1) - \frac{1}{2\left(1+(z/\alpha)^2\right)}\right]$$

eq. (1.18)

and,

$$\sigma_z = \frac{\sigma_{max}}{1+(z/\alpha)^2}$$

eq. (1.19)

Therefore, the maximum shear stresses developed are represented by equations (1.20) and (1.21) (Seely and Smith 1965; Timoshenko and Goodier 1970; Shigley and Mishke 2001)

$$\tau_{yz} = \tau_{xz} = \frac{1}{2}(\sigma_x - \sigma_z)$$

eq. (1.20)

and $\tau_{xy} = 0$

eq. (1.21)

The maximum shear stress is developed on the load axis, approximately 0.48α below the surface. Ductile materials first yield at the point of maximum shear stress (Timoshenko and Goodier 1970; Shigley and Mishke 2001). The values of stress components below the surface may be plotted as a function of maximum stress of contacting spheres.

The normal displacement (approach of distant points on the two bodies) is given by following expression.

$$\delta = \left(\frac{3F(1-v_1^2)}{4E_1}\right)^{2/3}\left(\frac{1}{R}\right)^{1/3}$$

eq. (1.22)

Quasi-static compression tests may be performed using a universal testing machine or a texture analyser. In case of small-sized materials, the material to be tested may be glued to the base plate (ASAE 1998). For example, an individual seed is loaded between two parallel plates and compressed until the seed fails (ASAE 1998; Saiedirad et al. 2008). The slow speed of the compression tool allows the material to be compressed for an appreciable time before failure occurs. The point of inflection may be determined visually from the force-deformation curve (Fig. 1.4) to compute apparent modulus of elasticity (Mohsenin 1986; ASAE 1998; Sayyah and Minaei 2004). For conducting compression tests, the procedures prescribed by ASAE (ASAE 1998) should be followed.

Factors Affecting Force Deformation Behaviour

Moisture content of the material plays a significant role in mechanical properties of food materials. Moisture would also greatly affect the stress-strain behaviour of dried food products such as spaghetti noodles or crackers. Such materials typically have moisture ranging from 5 to 30%, while fruit and vegetables have moisture contents of 75–90% (wet basis).

Strain rate also affects the stress-strain behaviour of agricultural materials and food products. More stress is usually required to produce a given amount of strain at higher strain rates. This is true for grains and seeds as well as dry food material. The behaviour of fruit and vegetable tissue is more complex. When the cells are ruptured, more stress is required to produce a given amount of strain at the faster loading rate. Strain rate has a relatively small effect at the intermediate water potential.

Compression of agricultural materials and food products usually produces a relatively large plastic strain. As a result, their stress-strain bahaviour changes under repeated or cyclic loading. Most of the plastic strain occurs during the first cycle of loading.

1.4.3 Stress Relaxation

If agricultural materials and food products are deformed to a fixed strain and the strain is held constant, the stress required to maintain the deformation decreases with time. This is called *stress relaxation*. For example, in the behaviour of a cylindrical sample of potato tissue at a strain of 10% the decrease in stress is extremely rapid during the first 5 or 10 seconds of loading. The initial stress is approximately 0.6 MPa, which decreases to 0.1 Mpa in just 2 seconds (Pitt 1984). The additional decrease during the next 18 minutes is relatively small. The stress as a function of time may be described by a sum of a constant and one or more exponential terms.

1.4.4 Creep

In bulk handling situations, such as when potatoes are placed in a pile or blocks of cheese are placed on the top of one another, a constant load is applied to agricultural materials and food products. If the stresses are relatively large, the material will continue to deform with time. This increase in strain is called creep. There is an almost instantaneous initial deformation followed by continual increases in strain as time of loading increases. However, the rate of change in stress with time (the slope of the curve) decreases exponentially with time. Eventually, the relationship between strain and time becomes nearly linear. This type of behavior is also typical of fruit and vegetables.

1.4.5 Deformation Testing Using Other Geometries

In many situations, agricultural materials are not loaded in simple compression. Although compression at the surface imposes compressive stresses in the vicinity of the applied load, other portions may be under tension. If fruit or vegetables absorb water rapidly, they may expand, producing tensile forces in the skin. When grains and seeds are dried, the outer layers may shrink producing tensile stresses in the layers and in the surrounding pericarp. For such agricultural materials and food products, the response under tensile loading may be very different from the response under compressive loading. Furthermore, compression, shear, and tensile forces occur in many loading situations and when evaluating failure (breaking apart), it may be desirable to test the response of the material to all three types of forces.

1.4.6 Tensile Loading

Tensile tests are used to measure the adhesion of a food to a surface. Adhesion is the force that resists the separation of the two bodies in contact. In this type of test, the sample of food has a probe pressed onto it after which the extraction force is measured. Important textural characteristics such as elasticity of spaghetti and extensibility of dough are some examples of tensile tests. Tensile tests have mainly been performed for meat analysis where breaking strength is the best parameter for predicting tenderness in cooked meat.

Tensile testing is less common than compression testing because it is more difficult to grip the sample in such a manner that a tensile load can be applied. If the ends of a bar of uniform cross-section are clamped, compression of the ends causes stress concentrations to develop, which promote failure in the vicinity of the clamp. This problem is usually

16 *Methods in Food Analysis*

overcome by making the sample wider in the vicinity of the clamp (Fig. 1.5A). Ideally, the sample will fail at its narrowest point. Therefore, stress is calculated from the minimum cross-sectional area. In some samples, a punch can be used to place a mark at two points on the sample and deformation (L) becomes the change in distance between the two points. A device called an extensometer can also be used to measure L. The extensometer is clamped to the sample at two points and it automatically registers the change in length on a dial gauge. The engineering strain can be calculated from L.

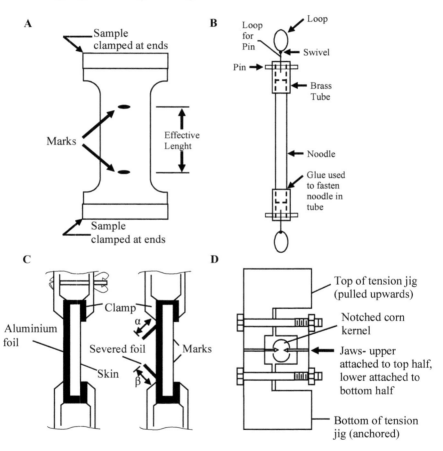

Figure 1.5 Methods for tensile testing of agricultural materials and food products. A. Typical sample shape used (The sample is wider at the ends where it is clamped into the testing device); B. Technique for tensile testing of noodles (Cummings 1981); C. Technique for testing tomato skins (Murase and Merva 1977); D. Apparatus used to conduct tensile tests on a single corn kernel (Ekstrom et al. 1966).

Some of the most frequent applications of tensile testing are testing of pericarp of grains and seeds or the epidermis of fruit and vegetables in which the sample is cut into the shape shown in Fig.1.5A (Liu et al. 1989 for soybean seed coats). Cummings and Okos (1983) tested noodles using rapid-setting glue to secure them in a metal tube slightly larger in diameter than the noodles (Fig. 1.5B). The tubes were attached to a tensile testing machine with wires and loops which allowed the sample to align itself with the line of application of the force. Murase and Merva (1977) wrapped tomato skin samples in aluminium foil, clamped the ends, severed the aluminium foil to expose the sample, and then soaked the sample in solutions of known water potential while applying a tensile force (Fig. 1.5C). Notches may also be made in a corn kernel and tensile force applied with a special supporting jig (Fig. 1.5D) (Ekstrom et al. 1966).

1.4.7 Fracture Test

Fracturability is the parameter that was initially called "brittleness". It is the force with which a sample crumbles, cracks or shatters. Foods exhibiting fracturability are products that possess a high degree of hardness and low degree of adhesiveness. The degree of fracturability of a food is measured as the horizontal force with which a food moves away from the point where the vertical force is applied. Another factor that helps in determining fracturability is the suddenness with which the food breaks.

1.4.8 Cutting and Shearing Test

There are many single-blade or multi-blade fixtures available with universal testing machines or texture analysers that cut or shear through the food samples. The maximum force required and the work done is taken as an index of firmness, toughness or fibrousness of the sample. Although the term "shear" is used to describe the action of such fixtures, both compression and tension forces are developed as well. Cutting and shearing test is usually done for foods with a fibrous structure, which includes meat, meat products and vegetables.

1.4.9 Bending and Snapping Test

Bending is a combination of compression, tension and shear. Snap test, defined as breaking suddenly upon the application of a force, is a desirable textural property in most crisp foods, such as fresh green beans and other vegetables, potato chips and other snack items. The ability to snap is a measure of the temper of chocolate, the moisture content of crisp cookies,

18 Methods in Food Analysis

the turgor of fresh vegetables and the amounts of shortening in baked goods. The sharp cracking sound that usually accompanies snapping is the result of high-energy sound waves generated when the stressed material fractures rapidly and the broken parts return to their former configuration.

Bending tests are performed by cutting a sample in the shape of a beam and placing it on a stand which supports it at two points separated by a distance L (Fig. 1.6). If a known force is applied to the centre of the beam, the modulus of elasticity may be determined from the deflection at the point of application of force. E may be calculated from the formula for deflection of a simply supported beam having a cross-section of height h and width b. Assuming that the force F is applied in the direction h so that the neutral axis of the beam is at $h/2$, and that the beam cross-section has the moment of inertia I, about the neutral axis, the value of E is given by following equation.

$$E = \frac{FL^3}{48DI}$$
eq. (1.23)

Where $I = \dfrac{bh^3}{12}$
eq. (1.24)

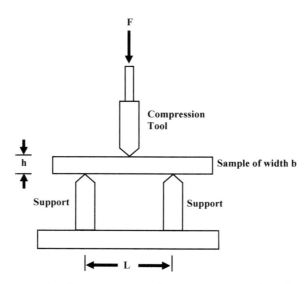

Figure 1.6 Three-point bending test on a sample cut into a rectangular cuboid. The sample is "simply supported" by the edges of the two supports and compressed by the edge of the compression tool attached to a movable cross-head. Compression tool applies force midway between the two supports.

Eq. (1.23) is only valid when E in compression is equal to E in tension. Therefore, the bending test may not be appropriate for the determination of E in many food materials.

The bending test may be used to determine the critical tensile stress at failure. For a simply supported beam with a force F applied halfway between the supports, the maximum tensile stress, σ_{max}, occurs at the bottom surface of the beam (the surface opposite to that on which the force is applied). The neutral axis is the plane where $\sigma=0$. If c is the distance from the surface to the neutral axis ($c=h/2$ for the simply supported beam) and M is moment about the neutral axis, then

$$\sigma_{max} = \frac{M_c}{I} = \frac{3FL}{2bh^2} \qquad \text{eq. (1.25)}$$

1.4.10 Puncture and Penetration Test

In a puncture or penetration test the probe penetrates into the test sample by a combination of compression and shear forces that cause irreversible changes in the sample. The puncture test measures the force required to reach a specified depth, whereas penetration test measures the depth of penetration under a constant load. In this test, the force necessary to achieve a certain penetration depth is measured and used as a measure of hardness, firmness or toughness. Puncture and penetration tests are commonly used in the testing of fresh fruit and vegetables, cheese, confectionery and the spreadability of butter and margarine. Penetration tests, such as the Bloom test, have also been used extensively for testing the rigidity of gels.

1.4.11 Texture Profile Analysis (TPA)

Texture profile analysis is also known as the two bite test. A number of product characteristics may be quantified and standard methods have been established to evaluate parameters such as adhesiveness, cohesiveness and springiness of food products. This test is usually performed in compression. In this test the specimen is compressed to the point where material reaches the bioyield point (first bite) and then the force is removed gradually. After a relaxation time, the force is again applied (second bite) till material fails. Typical texture profile analysis curve for pears is shown in Fig. 1.7.

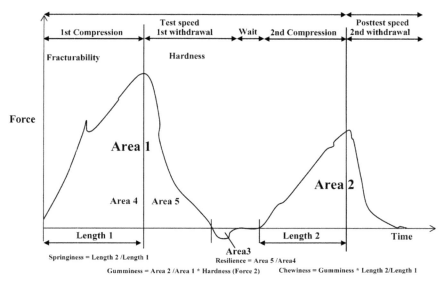

Figure 1.7 Typical texture profile analysis curve.

The various parameters determined from the TPA tests are described below.

Hardness: It is the force necessary to attain deformation; given as the final peak of the TPA curve (Fig. 1.6), which is the force value corresponding to the first major peak (the maximum force during the first cycle of compression). It is also known as firmness.

Fracturability: It is the force at which the material fractures (height of the first significant break in the peak of TPA curve). A sample with a high degree of hardness and low cohesiveness will fracture. This is also called brittleness. Fracturability is the force value corresponding to the fracture peak (if there is one).

Springness: It is the height that the food recovers during the time that elapses between the end of the first cycle and the start of the second cycle. The rate at which a deformed sample goes back to its undeformed condition after deforming force is important. This is also called elasticity.

Stringness: It is defined as the distance that the product is extended during de-compression before separating from the probe.

Adhesiveness: It is the measure of the work necessary to overcome the attractive forces between the surface of the sample and surface of the probe with which the sample comes into contact. If adhesiveness is larger than cohesiveness, then part of the sample will adhere to the probe.

Cohessiveness: It is the measure of the strength of internal bonds making up the body of the sample. If adhesiveness is smaller than cohesiveness, then the probe will remain clean as the product has the ability to hold together.

Chewiness: It is the measure of the energy required to masticate a semi-solid sample to a steady-state of swallowing (Hardness × Cohesiveness × Adhesiveness).

Gumminess: It is the measure of the energy required to disintegrate a semi-solid sample to a steady state of swallowing (Hardness × Cohesiveness).

1.4.12 Torsional Loading

The torsion test is a method well-suited for determining the failure properties of fruit, vegetables and other food products such as protein gels (Diehl et al. 1979; Hamann 1983). It is particularly useful because the shear and normal stresses are equal during loading and therefore the plane of failure indicates which of these loadings is most likely to cause failure.

In the torsion test, undesirable stress concentrations develop at the points of application of the torque at the ends of the sample. Therefore, a sample of varying diameter is used. The values of normal stresses/strains and shear stresses/strains may be calculated from the formulas summarized by Hamann (1983). The choice of formula depends on the assumptions made about the relationship between the applied moment, M and the angle of twist, Ψ. If M is assumed to be a linear function of Ψ all the way to the point of failure, then the values for the maximum shear stress, τ_{max} and maximum shear strain γ_{max}, are given by:

$$\tau_{max} = \frac{2KM}{\pi r_{min}^3} \qquad \text{eq. (1.26)}$$

$$\gamma_{max} = \frac{2K\psi}{\pi r_{min}^3 Q} \qquad \text{eq. (1.27)}$$

1.4.13 Test Specimen and Testing Conditions

Determination of mechanical properties of food materials is a technical job in which precision plays an important role. The following points must be considered when testing food materials.

- Specimens should be tested in their original size and shape.
- The specimen may be tested under 3 different conditions: (1) fresh, (2) frozen and thawed, or (3) cooked and dried.
- Tests on fresh specimens must be conducted before the time of exposure to air exceeds 10 min in order to avoid changes caused by drying of the specimen.
- Frozen specimens must be thawed, brought to room temperature (22 ± 2 °C), and tested before drying occurs.
- Cooked specimens should be air-dried for about 24 hours at room temperature before testing.
- Because of the large variance inherent in specimens, each experiment must be statistically designed to have enough test specimens for an acceptable level of confidence in the results. A minimum of 25 specimens should be used.
- For shear tests, a crosshead speed of 5 mm/min should be used.
- For the bending test, a crosshead speed of 10 mm/min should be used.

1.5 Steady State Rheology

Steady state relationship between shear stress-shear rate of food materials is expressed in terms of the power law model or the Herschel-Bulkley model. The Herschel-Bulkley model is used for yield stress fluids. Yield stress fluids behave like a solid until a minimum stress, known as yield stress, is overcome for beginning of the flow of the material (Fig. 1.1).

Power law model: $\tau = K\dot{\gamma}^n$ eq. (1.28)

Herschel-Bulkley model: $\tau = \tau_o + K\dot{\gamma}^n$ eq. (1.29)

where τ is shear stress (Pa), τ_o is yield stress (Pa), $\dot{\gamma}$ is shear rate (s^{-1}), K is consistency index (Pa.s^n) and n is flow behaviour index (dimensionless) expressing the extent of deviation from Newtonian behaviour.

Dependence of the flow behaviour of fluid foods on temperature can be described by the Arrhenius relationship (Saravacos 1970; Rao 1986; Steffe 1996):

$$K = A_K \exp(E_K / RT)$$ eq. (1.30)

where A_K is frequency factor (Pa.s^n), E_K represents activation energy (J/mol), R is gas law constant (R = 8.314 J/mol K), and T is absolute temperature (K).

Yield stress is the point when the shear stress-shear rate curve starts showing deviation of shear rate from zero. This condition indicates initiation of flow and the corresponding shear stress is taken as the yield stress.

1.5.1 Time Dependent Rheology

Time dependent shear stress decay characteristics have been mathematically described by several researchers (Weltman 1943; Hahn et al. 1959; Figoni and Shoemaker 1983; Nguyen et al. 1998). The Weltman model (1943) assumed logarithmic decay of shear stress in the absence of any equilibrium condition. The Weltman model was later modified by Hahn et al. (1959) to include an equilibrium shear stress term. Figoni and Shoemaker (1983) described the stress decay process with a first-order kinetic model with a non-zero equilibrium value. The structural kinetic model (Nguyen et al. 1998; Abu-Jdayil 2003) postulates that the change in the rheological properties is associated with shear-induced breakdown of the internal fluid structure. To quantify the time dependence of mango jam at selected shear rates and temperatures, shear stress and time of shearing data were fitted to the Weltman, Hahn, Figoni and Shoemaker, and structural kinetic models.

Weltman Model

Weltman model (1943) is expressed as

$$\tau = A - B \ln t \qquad \text{eq. (1.31)}$$

where τ is shear stress (Pa) at any given time of shearing (t). The parameter A represents the initial stress while B is time coefficient of structure breakdown.

Hahn Model

Hahn et al. (1959) evaluated the Weltman model and found plots of τ versus $ln(t)$ for the mineral oil to be sigmoidal but not linear. They argued on theoretical basis that stress decay of thixotropic substances should instead follow the first-order type relationship.

$$\log(\tau - \tau_e) = P - at \qquad \text{eq. (1.32)}$$

where τ_e is the equilibrium shear stress value, which is reached after a long shearing time; P represents the initial shear stress and a indicates the rate of structural breakdown for the sample.

Figoni and Shoemaker Model

Figoni and Shoemaker (1983) suggested a thixotropic model based on their work on transient rheology of mayonnaise.

$$\tau = \tau_e + (\tau_{max} - \tau_e)\exp(-kt) \qquad \text{eq. (1.33)}$$

where, τ_{max} is initial shear stress; $(\tau_{max}-\tau_e)$ represents the quantity of breakdown structure for shearing; and k is a kinetic constant of structural breakdown.

Structural kinetic model

The observed time-dependent flow behaviour of the food materials is also modelled using the structural kinetic approach (Nguyen et al. 1998; Abu-Jdayil 2003). This model postulates that the change in the rheological properties is associated with shear-induced breakdown of the internal fluid structure in the food. Using the analogy with chemical reactions, the structural breakdown process may be expressed as breakdown from structured to non-structured. The rate of breakdown of the structure during shear depends on the kinetics of the above reaction. Based on the experimental results from the transient measurements at constant shear rates, and from the step change in shear rate measurements, it may be assumed that the thixotropic structure in food breaks down irreversibly without significant build-up.

Let $\Psi = \Psi(\gamma, t)$ be a dimensionless parameter representing the structured state at any time t and under an applied shear rate γ. The rate of structural breakdown may be expressed as

$$-\frac{d\Psi}{dt} = k(\Psi - \Psi_\alpha)^m \qquad \text{eq. (1.34)}$$

where $k = k(\gamma)$ is rate constant; α is function of shear rate (γ); and m is the order of the breakdown 'reaction'. Initially, at the fully structured state, $t = 0$; $\Psi = \Psi_o$; and at steady state $t = t$: $\Psi = \Psi_\alpha$. At a constant applied shear rate, integration of equation (35) from initial time ($t = 0$) to a time (t) yields

$$(\Psi - \Psi_\alpha)^{1-m} = (m-1)kt + (\Psi_o - \Psi_\alpha)^{1-m} \qquad \text{eq. (1.35)}$$

To apply equation (1.35) to the experimental transient viscosity data, a relationship between Ψ and measurable rheological quantities needs to be determined. Ψ may be defined in terms of the apparent viscosity (η) as:

$$\Psi(\gamma,t) = \frac{\eta - \eta_\alpha}{\eta_o - \eta_\alpha} \qquad \text{eq. (1.36)}$$

where, η_o is initial apparent viscosity at $t = 0$ (structured state); η is the apparent viscosity at time t; and, η_α is the final or equilibrium apparent viscosity at $t \to \alpha$ (equilibrium structured state). Both η_o and η_α are functions of the applied shear rate only.

Substituting equation (1.36) into equation (1.35) gives the expression, for a fixed shear rate:

$$(\eta - \eta_\alpha)^{1-m} = (m-1)kt + (\eta_o - \eta_\alpha)^{1-m} \qquad \text{eq. (1.37)}$$

The form of equation (1.37) allows a simple way for testing the validity of the model and determination of the model parameters m and k. Equation (1.37) is valid only under the constant shear rate condition (Nguyen et al. 1998).

1.6 Viscoelasticity

Measurement of viscoelastic properties of food materials may be carried out using dynamic oscillation of shear stress or strain (Steffe 1996; Gunasekaran and Ak 2000). Harmonic oscillation of shear stress involves a material being oscillated sinusoidally with varying stress while the resulting strain is measured (Zhong 2003). Dynamic tests are carried out based on four assumptions: (i) a constant stress or strain throughout the sample; (ii) no slip of the sample; (iii) sample homogeneity; and (iv) measurements are performed within the linear viscoelastic region. Further, key parameters like storage modulus, loss modulus, complex modulus, and phase angle generated during dynamic oscillation studies describe the viscoelastic behaviour of a material.

Storage and Loss Modulus

The storage modulus (G') indicates the degree of elastic behaviour in a material. Shear storage modulus is the component in phase with the strain, or the elastic behaviour.

$$G' = G.\cos\delta \qquad \text{eq. (1.38)}$$

Loss modulus (G') is the component out of phase with the strain or viscous behaviour. Thus shear loss modulus is an indication of the viscous properties of the material.

$$G'' = G.\sin\delta \qquad \text{eq. (1.39)}$$

Complex Modulus

The complex shear modulus (G^*) includes both storage and loss moduli values and is an indicator of the strength of a gel.

26 Methods in Food Analysis

$$G^* = (G'^2 + G''^2)^{0.5} \qquad \text{eq. (1.40)}$$

Phase Angle

Phase angle (δ) is directly related to the energy lost per cycle divided by the energy stored per cycle (Steffe 1996). Phase angles can vary from 0 to 90°, with 0° indicating an ideal solid material (Hookean solid) and 90° indicating an ideal viscous material (Newtonian fluid). Phase angle is expressed as:

$$\tan(\delta) = \frac{G''}{G'} \qquad \text{eq. (1.41)}$$

1.6.1 Dynamic Rheology

The dynamic rheological test is performed by applying a small sinusoidal strain (or stress) and measuring the resulting stress (or strain). These small-amplitude oscillatory tests are commonly performed in shear and hence the abbreviation SAOS (small amplitude oscillatory shear; often at 1–3 or 5% (Gunasekaran and Ak 2000)) is commonly used to represent dynamic viscoelastic tests. This is to assure that the material response is in the linear range, i.e., the range within which the stress is proportional to the applied strain. Though not yet very popular, the dynamic tests are also performed in compression/tensile mode and in large amplitude shear mode (at strains in excess of 100%). The SAOS measurement is a non-destructive technique to get an insight of the material characteristics of any food material.

Strain Sweep Test

The domain of linear viscoelasticity is established by the oscillatory strain sweep experiment. Here, the strain amplitude of the oscillatory shear flow at a fixed frequency is continuously increased until the dynamic properties (storage- and loss-moduli) change significantly with strain. Below this strain level, viscoelastic response is linear. Strain sweeps at different frequency are performed for all the samples at selected temperatures to determine the linear viscoelasticity zone.

Frequency Sweep Test

After obtaining the value of strain, for which the material exhibited linear behaviour, the frequency sweep tests are performed at specified strain level over a frequency range. The oscillatory rheological parameters obtained are:

storage modulus (G′), loss modulus (G″), complex modulus (G*), complex dynamic viscosity (η^*), and loss angle (tan δ).

Shear Sweep Test

The shear sweep test is carried out over a selected range of the shear rate values. The apparent viscosity (η) against shear rate ($\dot{\gamma}$) data can be used with complex viscosity (η^*) against frequency (ω) data to test the validity of the Cox-Merz rule.

1.6.2 Analysis of Dynamic Rheological Data

The power law describes the rheological behaviour of an incipient gel within the linear viscoelastic region because the frequency dispersions of the dynamic mechanical spectra (G′ and G″) are more or less straight lines with different slopes. Therefore, each set of data can be fitted by power law equations as follows:

$$G' = a \times w^b \qquad \text{eq. (1.42)}$$

$$G'' = c \times w^d \qquad \text{eq. (1.43)}$$

where a is the low-frequency storage modulus (Pa); b is the power law index for storage modulus (dimensionless); c is the low-frequency loss modulus (Pa); and d is the power law index for the loss modulus (dimensionless).

Cox-Merz Rule

Several empirical relations have been proposed to relate the viscometric functions to linear viscoelastic properties. The Cox–Merz rule is one such relation:

$$|\eta^*(\omega)| = \eta(\dot{\gamma}) \text{ for } \omega = \dot{\gamma} \qquad \text{eq. (1.44)}$$

The Cox–Merz rule is a simple relationship that predicts whether the complex viscosity |$\eta^*(\omega)$| and steady shear viscosity $\eta(\dot{\gamma})$ are equivalent when the angular frequency (ω) is equal to the steady shear rate ($\dot{\gamma}$).

Although originally developed for synthetic polymers (Steffe 1996), the Cox-Merz rule and its modified forms have been applied to many liquid and semisolid foods (Bistany and Kokini 1983; Yu and Gunasekaran 2001). Deviation from the Cox-Merz relation has been attributed to various structural aspects and to particle-particle interactions, especially in highly concentrated systems (Yu and Gunasekaran 2001; Gleissle and Hochstein

2003). Compared to synthetic polymers, rheological behaviour of food materials may deviate from the Cox-Merz relation to a large extent. However, in many cases, it has been found that the foods follow the same general behaviour when a shift factor, A, is introduced (Bistany and Kokini 1983; Yu and Gunasekaran 2001; Gleissle and Hochstein 2003).

$$\left| \eta^*(\omega) \right| = \eta(A\dot{\gamma}) \Big|_{\omega=\dot{\gamma}}$$

eq. (1.45)

Another extended Cox-Merz rule (Bistany and Kokini 1983; Steffe 1996) has also been established for some food materials (condensed milk, tomato paste, ketchup, mayonnaise, wheat flour dough, and starch dispersions).

$$\eta(\dot{\gamma}) = K' \eta^*(\omega)^\alpha \Big|_{\omega=\dot{\gamma}}$$

eq. (1.46)

where K' and α are constants. In most cases both η^* versus ω and η versus $\dot{\gamma}$ may be approximated by a power law. Thus, when $\alpha = 1$, this relation reduces to the modified Cox-Merz rule.

Weak Gel Model

Gabriele et al. (2001) were the first to conceptualize food as a critical weak gel and proposed a power law relaxation modulus to describe the rheological behaviour of dough, jam, and yoghurt. This descriptive framework has also been used recently to explore other types of deformation including creep relaxation, and uniaxial and biaxial extension (Gabriele et al. 2004; Ng and Mckinley 2008). The so called 'weak gel model' (Gabriele et al. 2001) is extremely attractive because of its relative functional simplicity.

This model provides a direct link between the microstructure of the material and its rheological properties. The most important parameter introduced is the 'coordination number', z, which is the number of flow units interacting with each other to give the observed flow response. Above the Newtonian region, there exists a regime characterized by the following flow equation:

$$\left| G^* \right| = \sqrt{G'(\omega)^2 + G''(\omega)^2} = A\omega^{1/z}$$

eq. (1.47)

where, A is a constant that may be interpreted as the 'interaction strength' between the flow rheological units. Thus, material functions of food system in the linear viscoelastic regime may be well described by only two parameters (A and z).

Creep Test

In creep test, a constant shear stress in the linear viscoelastic region (obtained from strain-sweep tests) is applied and the resultant strain is measured as a function of time (t). After time t, the stress is removed and the recovery of strain is measured for $2t$. The results are expressed in terms of instantaneous shear creep compliance (J_t).

$$J_t = \frac{\gamma}{\tau} \qquad \text{eq. (1.48)}$$

The J_t versus time data during creep may be analyzed using the Peleg, Burgers, or Kelvin-Voigt models.

Peleg Model

Peleg (1980) suggested that creep data may be modelled by a linear relation:

$$\frac{t}{J_t} = k_1 + k_2 t \qquad \text{eq. (1.49)}$$

where, k_1 (Pa.s) and k_2 (Pa) are constants; $(1/k_1)$ represents the initial decay rate; and $(1/k_2)$ is the asymptotic strain.

Burgers Model

The Burgers model (Steffe 1996) in terms of compliance function is described by a four-parametric equation:

$$J_t = f(t) = J_o + J_1(1 - \exp(-\frac{t}{\lambda_{ret}})) + \frac{t}{\mu_0} \qquad \text{eq. (1.50)}$$

where J_o is the instantaneous compliance resulting from the instantaneous stress applied; J_1 is the time-dependent retarded elastic compliance (Kelvin component); λ_{ret} is the retardation time of the Kelvin component; and μ_o is the asymptotic viscosity of the material (viscous flow of the bond-free constituents). The instantaneous compliance (J_o) is the compliance at time zero and is determined by extrapolation of the compliance to zero time.

Kelvin Model

The Kelvin model (Steffe 1996) is a six-parameter mechanistic model, being an extension of the Burgers model:

30 Methods in Food Analysis

$$J_t = f(t) = J_o + J_1\left(1-\exp\left(-\frac{t}{(\lambda_{ret})_1}\right)\right) + J_2\left(1-\exp\left(-\frac{t}{(\lambda_{ret})_2}\right)\right) + \frac{t}{\mu_0} \quad \text{eq. (1.51)}$$

where $(\lambda_{ret})_1 = \frac{\mu_1}{G_1}$; and, $(\lambda_{ret})_2 = \frac{\mu_2}{G_2}$; J_1 and J_2 are the retarded compliance of first and second Voigt-Kelvin element; and $(\lambda_{ret})_1$ and $(\lambda_{ret})_2$ are the retardation times of the first and the second Voigt-Kelvin elements, respectively.

1.6.3 Gel Strength and Relaxation Exponent

The gel strength (S) and the relaxation exponent (n) are computed from the creep test data using the Winter and Chambon (1986) equation:

$$G(t) = S.t^{-n} \quad \text{eq. (1.52)}$$

where S (Pa.s^n) is the gel strength, which depends on the cross-linking density and the molecular chain flexibility; and n is related to the molecular structure and connectivity of the incipient gel (Nijenhuis 1997). Exponent n characterizes the critical gel: at $n < 0.5$, the cross-linker is in excess, and the opposite holds at $n > 0.5$ (Nijenhuis 1997). $G(t)$ is the relaxation function, which is established from creep test data as the reciprocal of J_t (Ferry 1980).

1.7 Rheometery

Rotational rheometers may operate in two modes: steady-shear or oscillatory. Steady shear is a condition in which the sheared fluid velocity, contained between the boundaries, remains constant at any single position. Furthermore, the velocity gradient across the fluid is a constant. Two test fixtures that are often used in steady-shear rotational viscometers are concentric cylinder and the cone-and-plate.

1.7.1 Cone and Plate Viscometers

In this type of viscometers, the fluid is held by its own surface tension between a cone of small angle that just touches a flat surface (Fig. 1.8A). The torque generated due to the drag of the fluid on the cone is measured as one of the members rotates while the other remains stationary. Test quality is best when the cone angle (θ) is small, and large errors may be encountered when the gap is improperly set or not well maintained.

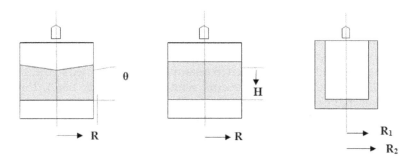

A. Cone and plate   B. Plate and plate   C. Concentric cylinder

Figure 1.8 Types of viscometers.

The shear stress may be determined for a cone-and-plate configuration as:

$$v = \frac{3M}{2\pi R^3} \qquad \text{eq. (1.53)}$$

while the shear rate is calculated as:

$$\dot{\gamma} = \frac{r\Omega}{r\tan\theta} = \frac{\Omega}{\tan\theta} \qquad \text{eq. (1.54)}$$

Some of the merits of the cone-and-plate method include (i) small sample size (roughly one gram), (ii) uniform shear rate and (iii) easy to load the sample and clean the instrument.

1.7.2 Plate and Plate Viscometers

The measuring geometry of this viscometer consists of two parallel plates at a distance H from each other (Fig. 1.8B), of which one can rotate. This measuring geometry is especially suited for the study of materials whose structure is not to be disturbed at all or as little as possible while one is setting up the rheometer, for example, thixotropic materials or products with a yield stress, which may take a long time to recover. The advantages of the plate-and-plate viscometers include (i) small sample size (roughly one gram), (ii) temperature variable without reloading the sample, and (iii) easy to load and clean. A limitation of this method is that shear rate is not uniform and so it is useful for linear viscoelasticity, but not suited for nonlinear studies.

1.7.3 Concentric Cylinders

This rheological attachment consists of a cylindrical fixture shape, commonly called a bob, with a radius R_b, suspended from a torque (M) measuring device that is immersed in a sample fluid contained in a slightly larger cylinder, referred to as the cup, with radius R_c (Fig. 1.8C). Torque is an action that generates rotation about an axis and is the product of a force and the perpendicular distance to the axis of rotation (l), called the moment arm.

$$v_b = \frac{M}{2\pi h R_b^2} \qquad \text{eq. (1.55)}$$

Concentric cylinders are advantageous for measuring the rheological properties because (i) even for large radii the shear rate is nearly constant, and (ii) it is ideally suited for pourable liquids.

There are some empirical viscometers in which a paddle, a cylinder, or a bar rotates in a container, usually with large clearances between the rotating member and the wall. The geometry of these viscometers is complex and usually not amenable to rigorous mathematical analysis. These instruments are generally rugged, moderate in cost, and are fairly easy to operate. They are widely used in industry. Examples of this type of viscometers include Brookfield viscometers, and FMC consistometer. The main advantages of a rotational rheometer is that this equipment allows a continuous measuring of the relationship between the shear rate and the shear stress, also allowing the analysis of time-dependent fluids. Such equipment maintains a constant rotational speed corresponding to a certain shear rate; the shear stress is obtained through torque measurement.

1.8 Rheology of Fruit and Vegetable Products

The rheological properties of fruit and vegetables are of interest to plant physiologists, horticulturists, and agricultural/food engineers due to different reasons. Fruit, vegetables and their value-added products are a part of contemporary human diet now. Rheological properties are of importance in several aspects of production, handling, processing, manufacturing, human perception of product quality, etc. (Steffe 1996). The rheology of some of the products manufactured from fruit and vegetables is described in the following section.

1.8.1 Fruit Juices

Fruit juices are valuable semi-finished products for use in the production of beverages and powders. The conventional mode of fruit processing and preservation is in the form of juices or pulps (purees). However, preservation of juices is not economical due to the existence of high water content (75 to 90%).

It is possible to classify the juices into three groups: 1) clarified and depectinated concentrates; 2) clarified and non-depectinated concentrates; and 3) concentrates with suspended solids (Ibarz et al. 1996b). Generally, the first of these product groups presents a Newtonian behaviour. The presence of pectin substances or/and suspended solid particles causes non-Newtonian behaviour of juices and concentrates. Therefore, in order to describe the flow behaviour of these products, the power law, Herschley-Bulkley or Bingham models are used.

A good knowledge of the rheological behaviour of juices is fundamental to understand and improve the technological processes. In this matter, rheology has application in process equipment design, product development, storage and transportation and quality control of juices (Oomah et al. 1999). The rheological behaviour of fruit juices and concentrates is influenced by their composition, especially type of fruit and the treatments performed in the technological process (Ibarz et al. 1996a). In addition, factors such as temperature and concentration influence the rheological properties of these products.

Suárez-Quintanilla et al. (2003) analyzed rheological behaviour of three fruit fluids, previous to and after they were concentrated in a vacuum evaporator. The three fruits (papaya, melon and watermelon) were physico-chemically characterized (acidity, Brix, color, density, pH, solids and water contents) to identify their ripeness and particular stage. Fresh juices and concentrates were rheologically analyzed with two mathematical models to fit the flow data: power law and Herschel-Bulkley relationships. All samples (fresh and concentrated) exhibited non-Newtonian behaviour. The solids concentration mainly affected two rheological properties, the consistency coefficient and the yield stress, being more noticeable on the consistency coefficient. An exponential relationship was obtained for consistency coefficient as a function of solids content. The flow behaviour index was basically constant at the four studied levels of solids content.

The rheological measurements of cherry juice concentrate samples were carried out with a rotating rheometer Rheolab MC1 (Physica, Germany) measuring system with coaxial cylinders (cup diameter: 48 mm, bob diameter: 45 mm) by Juszczak and Fortuna (2004). The flow curves at different temperatures (10, 20, 30, 40, 50 and 60 °C) were obtained in the range of shear rates of 1–300 s^{-1}. The results indicated that concentrated cherry juices exhibit Newtonian behaviour and were strongly dependent on temperature and the soluble solid content. The effect of temperature on viscosity was described by the Arrhenius equation while the values of flow activation energy ranged from 24.68 to 38.48 kJ/mol.

Kaya and Belibağli (2002) studied rheological characterisation of pekmez (common grape product in Turkey). Pekmez is a concentrated and shelf-life extended form of grape juice, formed by boiling without the addition of sugar or other food additives. The rheological behaviour of solid pekmez (concentrated grape juice, °Brix=82.1) with different solid contents (72.9, 66.8, 57.2 and 52.1 °Brix) was studied at the temperatures of 10, 20, 30, 40 and 50 °C using a controlled-stress rheometer. Solid pekmez exhibited non-Newtonian behaviour. However, the diluted samples were Newtonian and the effect of temperature on viscosity was described by means of the Arrhenius equation.

Rheological measurements of tomato juice were carried out in a Haake RS 80 rheometer with controlled stress (σ), using a Couette geometry (concentric cylinder; Haake Z40-DIN) by Augusto et al. (2011). The cup and bob radius ratio was 1.0847 (bob radius = 20 mm). The temperature was maintained constant at 25 °C by using a water-bath (Phoenix Thermo Haake C25P) with deviations below ±0.3 °C. Tomato juice showed weak gel behaviour, with storage modulus higher than loss modulus at the evaluated frequencies.

1.8.2 Jams

Manufacturing of traditional jams involves use of fruit, sugar, pectin and organic acids such as citric acid. In traditional products high soluble solids content is desired to increase the product shelf life, achieving storage and transportation in ambient conditions. The high content of soluble solids is achieved by adding sugar to around 55%. The rheological properties of the final jam are affected by the amount and type of sugar added, proportion and kind of gelling agent used, fruit content, and process temperature. The quality of the raw material and the manufacturing process are the indicators of the final product quality (Nindo et al. 2005).

Fruit products that have fibre content or are high in pectin need less pectin in their jam formulation. Fruit with high pectin are usually difficult to handle in processing because of the matter protopectin being extracted. This will lead to changes in the setting behaviour, especially if the pectin produced has rapid setting behaviour. If the jam includes pulp particles or insoluble fibre they will affect the jam's texture and make it firmer, moreover, they will decrease jams tendency to syneresis, although they do not participate in the gelling system, this affect is due to their water-binding properties (Mizrahi 1979).

Alvarez et al. (2006) studied the rheological behaviour of selected jams at 20 to 40 °C in a rotational viscometer. The rheograms were fitted with power-law, Carreau, Carreau-Yasuda, Herschel-Bulkley, and Cross models and it was found that all of the models explained the rheological behaviour of jam. The jams exhibited pseudoplastic behaviour and the suspended solids influenced the consistency index.

Basu and Shivhare (2010) showed that mango jam behaved as a pseudoplastic fluid exhibiting yield stress. The Herschel-Bulkley model adequately described the steady-state rheological properties of mango jam. Time-dependent structural breakdown process was described well by the Hahn model.

1.8.3 Puree

Fruit purees are manufactured with varied fruit (peach, apple, pear and banana, or others like apricot, orange or pineapple), and may be enriched or not with vitamins. Water or juices are added to fruit (Alvarez et al. 2008). There are several steps in the preparation of fruit and vegetable purees: peeling of the skin of the fruit, size reduction, heating to either soften the tissue and/or inactivate enzymes, straining of the heated mass through finishers (finishing), and the addition of starch or sugar to obtain the desired consistency. The finishing operation (screen size and finisher speed) can be expected to have significant influence on the consistency of the puree due to its effect on the quantity and the dimensions of the pulp (Rao et al. 1986). Flow properties of fruit purees are of considerable interest in the development of fruit products for technological and marketing reasons. Together with others, they provide the information necessary for the optimum design of unit processes; contribute to the quality control in both manufacturing processes and final product; limit the acceptability and the field of application of a new product; and finally, they are a powerful tool for understanding molecular structure changes.

Balestraa et al. (2011) evaluated rheological behaviour of the purees (peach, apple and pear) and determined a possible correlation between the Bostwick consistency and the rheological parameters (such as viscosity and yield stress values) obtained by applying the Casson model to the flow curves obtained by using a rheometer. The viscosity variation against the shear rate of fruit purees was exponential and therefore they were non-Newtonian fluids. The Casson model was found to fit adequately over the entire shear-rate range.

El-Samahy et al. (2003) indicated that guava puree treated with Rohament PL enzyme had a non-Newtonian pseudoplastic behavior, with the presence of thixotropy. Increasing the enzyme concentration, holding time and temperature resulted in reduction of the consistency index (m) and the apparent viscosity. On the other hand, guava puree treated with pectinex ultra SP-L enzyme showed a Newtonian behaviour. Rheological measurements were carried out on the Brookfield viscometer model (RV) DV-II and the power law model was used to calculate the flow behaviour parameters. Concentrating the guava puree did not change its pseudoplastic time-dependent behaviour. Treating the puree with pectinex ultra SP-L enzyme before concentration caused a change in the flow behaviour of the resultant concentrate from non-Newtonian to Newtonian.

Espinosa et al. (2011) conducted steady-state measurements of apple purees using a concentric cylinder geometry consisting of a rotating inner cylinder (MCR-301 controlled stress rheometer, Anton Paar, Germany) at 20 °C in triplicate. A pre-shear rate of 43 s^{-1} was applied to the puree for 1 min, then to measure the apparent viscosity it was increased from 2.14 to 214 s^{-1} in 5 min and decreased from 214 to 2.14 s^{-1} in 5 min. The rheological properties were described by the Herschel-Bulkley model and the apple purees showed a shear thinning behaviour, presenting a yield stress. Increasing pulp content increased shear thinning behaviour and consistency index. Apparent viscosity and yield stress decreased with decreasing particle size and pulp content. Ditchfield et al. (2004) determined rheological behaviour of banana puree using a dynamic stress rheometer with a pressure coquette fixture, which allowed experiments to be conducted at high temperature. The shear stress values ranged from 10 to 170 Pa and the shear rate values from 105 to 103 s^{-1}. The Herschel-Bulkley model best fitted at all temperatures. Apparent viscosity decreased with increasing temperature, but increased with temperature from 50 to 60 °C and from 110 to 120 °C. The change in apparent viscosity could be due to the interactions of polysaccharides present in banana puree.

Rao and Palomino (1974) studied flow behaviour of banana puree using a tube viscometer at ambient temperature. The researchers concluded that banana puree is a pseudoplastic fluid and that the power law model was the best one to describe its behaviour. The power law model was also used for describing the rheological behaviour of banana puree at 26.5 °C by Holdsworth (1993).

Recently, Gundurao et al. (2011) determined effects of temperature (20, 40, 60, and 80 °C) and sugar syrup concentration (15, 20, 30, and 40° Brix) on thermo-physical and rheological properties of mango puree. Mango puree exhibited pseudo-plasticity during steady shear measurements between shear rates of 0.1 to 100 s^{-1}.

Alvarez et al. (2004) studied rheological behaviour of natural and commercial potato puree in steady and dynamic shear conditions at sample temperatures ranging from 25–65 °C. Both types of puree were frozen-then-thawed to study changes occurring in their structure that affected their rheological behaviour. All of the purees presented shear-thinning behaviour with yield stress. The effect of sample temperature on dynamic parameters was more significant in processed than in fresh purees. The Cox-Merz rule was not applicable to the steady and dynamic shear data on all of the purees studied. For a fresh commercial puree, the relationship between steady and dynamic shear data was non-linear. In both processed purees, the linearity or otherwise of the relationship between viscosities was dependent on sample temperature.

Ahmed et al. (2000) observed rheological characteristics of green chilli puree at 25 °C. Green chilli puree behaved as a shear-thinning fluid and the power law described well the shear stress–shear rate behaviour. The consistency index (K) decreased while the flow behaviour index (η) increased as the particle size of the puree decreased.

Coriander leaves have been used as a food flavouring agent in various cuisines since ancient times. Rheological characteristics of the coriander leaves puree was evaluated using a computer-controlled Haake rotational viscometer at 50, 60, 70 and 80 °C. The Herschel–Bulkley model adequately represented shear stress-shear rate data. Temperature dependency of the consistency index and apparent viscosity at the shear rate of 100 s^{-1} followed the Arrhenius relationship and the flow activation energy ranged between 17.2 and 17.9 kJ/mol (Ahmed et al. 2004).

1.8.4 Paste

Fruit paste is a product obtained in the same way as special non-gelified fruit marmalade but with lower water content (about 25% TSS in fruit paste). Lowering water content could be achieved by boiling or by drying the product using natural or artificial drying techniques. Rheological properties of date pulp (15 and 45°Brix) and concentrates (73°Brix) were analyzed by El-Samahy et al. (2006). They determined flow behaviour parameters based on different rheological models (power law, Bingham, and IPC paste) and also studied the effect of temperature on the rheological characteristics of date concentrate and date pulp, and calculated the activation energy based on the Arrhenius model.

Rheological parameters of date pastes were measured by a Bohlin Visco 88 viscometer (Bohlin instrument, UK) equipped with a cone-and-plate geometry (cone diameter 30 mm, cone angle 5°) and a heating circulator (Julabo, Germany) (Razavi and Karazhiyan 2012). The shear stress-shear rate data were then fitted using six models known as power law, Bingham plastic, Herschel–Bulkley, Casson, Sisko, and Vocaclo. The flow curves of date pastes were determined at four temperature levels of 20, 40, 60, and 70 °C by increasing the shear rate from 14 to 500 s^{-1}. The model that best fitted the experimental data at all temperatures was the Casson model. The viscosity decreased with increase of temperature.

Rheological characteristics (shear stress, shear rate and apparent viscosity) of ginger paste were measured using Brookfield RVDV-III rheometer in the temperature range of 25 to 65 °C (Ahmed 2004a). The paste showed pseudoplasticity with yield stress and shear stress-shear rate data were well described by the Herschel-Bulkley model. The yield stress decreased with increase in temperature in the range of 63.3–159.2 Pa. Ahmed (2004b) prepared ginger paste from fresh ginger by addition of 8% common salt and citric acid which was thermally processed and packed in glass, polyethyleneterephthalate or high-density-polyethylene containers and stored at 5±1 and 25±1 °C for 120 days. The rheological characteristics of the paste were studied by using a computer-controlled rotational viscometer over the temperature range of 20–80 °C. The ginger paste exhibited pseudoplasticity with yield stress and flow adequately described by the Herschel–Bulkley model. The yield stress decreased exponentially with process temperature and ranged between 3.86 and 27.82 Pa. The flow behaviour index (η) varied between 0.66 and 0.82 over the temperature range. Both consistency index and apparent viscosity decreased with increase in temperature and the process activation energies were in the range of 16.7 to 21.9 kJ/mol.

Fenugreek paste is a local food in Turkey, which is produced from pulverised fenugreek seeds and is used as an edible coating material in meat products. Isikli and Karababa (2005) studied flow behaviour and time-dependent flow properties of Fenugreek paste in the temperature range of 10–30 °C. Fenugreek paste exhibited non-Newtonian behaviour and the power law model adequately described the flow behaviour at different temperatures. It exhibited rheopectic behaviour that increased viscosity upon increasing the shear speed and the temperature.

Ahmed and Shivhare (2001) observed rheological behaviour and effects of packaging materials and storage temperatures on colour of garlic paste. Garlic paste behaved as a pseudoplastic material and the flow activation energy at 100 rpm was 13.30 kJ/mol.

1.8.5 Pulps

Fruit and vegetable pulps are obtained by mechanical treatment or, less often, by thermal treatment of fruit/vegetable followed by their preservation. Pulps may be classified as boiled or non-boiled. Factors affecting the rheological behaviour of the purees/pulps include total solids, total soluble solids, particle size, and temperature. It has been reported that fruit purees/pulps behave as non-Newtonian fluids (Holdsworth 1971).

Bhattacharya (1999) tested time-independent and time-dependent flow properties of mango pulp using a coaxial cylinder rheometer. Mango pulp was found to be a pseudoplastic liquid with yield stress, and exhibited thixotropic properties. The yield stress calculated using the Casson or Bingham plastic models had markedly higher values than those determined by stress relaxation, controlled stress experiments, or from stress-strain plots. The yield stress of mango pulp decreased as temperature increased.

Ahmed et al. (2005) studied the effect of high-pressure (HP) treatment (100–400 MPa for 15 or 30 min at 20 °C) on the rheological characteristics and colour of fresh and canned mango pulps. Differences were observed in the rheological behaviour of fresh and canned mango pulps treated with HP. Shear stress-shear rate data of pulps were well described by the Herschel–Bulkley model. The consistency index (K) of fresh pulp increased with pressure level from 100 to 200 MPa while a steady decrease was noticed for canned pulp.

The rheological behaviour of cabbage pulp was determined using a computer controlled rotational rheometer at temperatures of 20, 30, 40, and 50 °C by Gong et al. (2010). The flow behaviour of pulp was pseudoplastic and fitted well by the Casson model. The effect of soluble solids on the apparent viscosity was more pronounced than the effect of temperature.

1.9 Rheology, Texture and Product Quality

Texture is a very important parameter for sensory acceptance of fruit and vegetable products, depending largely on the composition of raw materials such as the type of fruit, fruit quantity and sugars used, additives like gelling agents, acidulates and the processing conditions. The quality criteria are decisively determined by the flavor, colour and consistency as well as the state of preservation and distribution of fruit/vegetables. Rheology has important role in food product development, quality control, sensory quality, design, and evaluation of the process equipment. Measurement of rheological parameters have been recognized as necessary to provide fundamental insights into the structural organization of food and important for understanding fluid flow and heat transfer. Rheological measurements allow the study of influence of chemical, mechanical and thermal treatments, the effects of additives, or the course of a curing reaction in the product characteristics. Krokida et al. (2001) assembled additional data on rheological properties of fluid fruit and vegetable puree products. They stated that fruit and vegetable juices/concentrates usually behave as non-Newtonian fluids, following the power law model. When fruit juices contain considerable amounts of pulps or are very concentrated, they may show an additional resistance to flow represented by a yield stress (Hernandez et al. 1995; Telis-Romero et al. 1999). Liquid foods have an extremely wide range in rheological properties. Classical tests play an important role in new product development and quality control of final goods. Thanks to improved rheometers, extended tests may be performed to obtain additional information, often without spending additional time.

1.10 Conclusion

Textural and rheological properties of fruit and vegetable products depend on a variety of factors including raw materials, additives used and the processing parameters. Knowledge of the textural and rheological properties of food products is essential for product development, design and evaluation of process equipment such as pumps, piping, heat exchangers, evaporators, sterilizers and mixers. Accurate measurement of textural and rheological properties helps the food manufacturer calculate volumetric flow rates, select pumps, determine pressure drops for pipe sizing and power consumption for pumping systems, and predict the heat transfer coefficients for heating, evaporation and sterilization.

References

Abu-Jdayil, B. 2003. Modeling the time-dependent rheological behavior of semisolid foodstuffs. Journal of Food Engineering. 57: 97–102.

Ahmed, J. and Shivhare, U.S. 2001. Thermal kinetics of color change, rheology, and storage characteristics of garlic puree/paste. Journal of Food Science. 66: 754–757.

Ahmed, J. 2004a. Rheological behaviour and colour changes of ginger paste during storage. International Journal of Food Science & Technology 39: 325–330.

Ahmed, J. 2004b. Effect of temperature on rheological characteristics of ginger paste. Emir. Journal Agricultural Scie. 16: 43–49.

Ahmed, J., Ramaswamy, H.S. and Hiremath, N. 2005. The effect of high pressure treatment on rheological characteristics and colour of mango pulp. International Journal of Food Science and Technology. 40: 885–895.

Ahmed, J., Shivhare, U.S. and Raghavan, G.S.V. 2000. Rheological characteristics and kinetics of colour degradation of green chili puree. Journal of Food Engineering. 44: 239–244.

Ahmed, J., Shivhare, U.S. and Singh, P. 2004. Colour kinetics and rheology of coriander leaf puree and storage characteristics of the paste. Food Chemistry. 84: 605–611.

Álvarez. E., Cancela, M.A., Delgado-Bastidas, N. and Maceiras, R. 2008. Rheological characterization of commercial baby fruit purees. International Journal of Food Properties. 11: 321–329.

Álvarez, E., Cancela, M.A. and Maceiras, R. 2006. Effect of temperature on rheological properties of different jams. International Journal of Food Properties. 9: 135–146.

Alvarez, M.D., Fernández, C. and Canet, W. 2004. Rheological behaviour of fresh and frozen potato puree in steady and dynamic shear at different temperatures. European Food Research and Technology. 218: 544–553.

ASAE. 1998. Compression test of food materials of convex shape. ASAE Standards S368.3. Michigan, USA.

Augusto, P.E.D., Falguera, V., Cristianini, M. and Ibarz, A. 2011. Viscoelastic properties of tomato juice. Procedia Food Science. 1: 589–593.

Balestraa, F., Coccia, E., Marsilioa, G. and Rosaa, M.D. 2011. Physico-chemical and rheological changes of fruit purees during storage. Procedia Food Science. 1: 576–582.

Basu, S. and Shivhare, U.S. 2010. Rheological, textural, microstructural and sensory properties of mango jam. Journal of Food Engineering. 100: 357–365.

Bhattacharya, S. 1999. Yield stress and time-dependent rheological properties of mango pulp. Journal of Food Science. 64: 1029–1033.

Bistany, K.L. and Kokini J.L. 1983. Dynamic viscoelastic properties of foods in texture control. Journal of Rheology. 27: 605–620.

Bourne, M.C. 1982. Food texture and viscosity. 1st ed. Academic Press, New York.
Bourne, M.C. 1993. Texture. 1st edn. pp. 4059–4065. *In*: Macrae, R., Robinson, R.K. and Sadler, M.J. (eds.). Encyclopaedia of Food Science, Food Technology and Nutrition. Academic Press, London.
Brusewitz, G.H., McCollum, T.G. and Zhang, X. 1991. Impact bruise resistance of peaches. Transactions of the ASAE. 34: 962–965.
Cummings, D.A. and Okos, M.R. 1983. Viscoelastic behaviour of extruded durum semolina as a function of temperature and moisture content. Transactions of ASAE. 26: 1883–1893.
Cummings, D.A. 1981. Modeling of stress development during drying of extruded durum semolina. Ph.D. Thesis, Department of Agricultural Engineering, Purdue University, West Lafayette, Indiana, USA.
Daubert, C.R. and Foegeding, E.A. 1998. Rheological Principles for Food Analysis. *In*: S.S. Nielsen (ed.). Food Analysis. 2nd edn., Aspen Publishers, Gaithersburg, U.S.A.
Diehl, K.C., Hamann, D.D. and Whitfield, J.K. 1979. Structural failure in selected raw fruits and vegetables. Journal of Texture Studies. 10: 371–400.
Ditchfield, C., Tadini, C.C., Singh, R. and Toledo, R.T. 2004. Rheological properties of banana puree at high temperatures, International Journal of Food Properties. 7: 571–584.
Ekstrom, G.A., Liljedhal, J.B. and Peart, R.M. 1966. Thermal expansion and tensile properties of corn kernels and their relationship to cracking during drying. Transactions of ASAE 9: 556–561.
El-Samahy, S.K., Askar, A.A., El-Mansy, H.A., Omran, H.T. and El-Salam, N.A. 2003. Flow behavior of guava puree as a function of different enzyme treatments and concentration. 3rd International Symposium on Food Rheology and Structure. 425–426.
El-Samahy, S.K., El-Hady, E.A., Mostafa, G.A. and Youssef, K.M. 2006. Rheological properties of date pulp and concentrate. Proceedings of the 4th International Symposium on Food Rheology and Structure. 583–584.
Espinosa, L., To, N., Symoneaux, R., Renard, C.M.G.C., Biaue, N. and Cuvelier, G. 2011. Effect of processing on rheological, structural and sensory properties of apple puree. Procedia Food Science. 1: 513–520.
Ferry, J.D. 1980. Viscoelastic Properties of Polymers. 3rd Edn., John Wiley and Sons, NewYork, USA.
Figoni, P.I. and Shoemaker, C.F. 1983. Characterization of time dependent flow properties of mayonnaise under steady shear. Journal of Texture Studies. 14: 431–442.
Gabriele, D., De Cindio, B. and D'Antona, P. 2001. A weak gel model for foods. Rheologica Acta. 40: 120–127.
Gabriele, D., Curcio, S., Migliori, M. and De Cindio, B. 2004. The use of rheology to characterize flow behavior of liquorice solutions. Journal of Food Process Engineering. 27: 464–475.
Gleissle, W. and Hochstein, B. 2003. Validity of the Cox-Merz rule for concentrated suspensions. Journal of Rheology. 47: 897–910.
Gong, Z., Zhang, M., Bhandari, B., Mujumdar, A.S. and Jin-Cai, S. 2010. Rheological properties of cabbage pulp. International Journal of Food Properties. 13: 1066–1073.
Gunasekaran, S. and Ak, M.M. 2000. Dynamic oscillatory shear testing of foods–selected applications. Trends in Food Science and Technology. 11: 115–127.
Gundurao, A., Ramaswamy, H.S. and Ahmed, J. 2011. Effect of solids concentration and temperature on thermo-physical and rheological properties of mango puree. International Journal of Food Properties. 14: 1018–1036.
Hahn, S.L., Ree, T. and Eyring, H. 1959. Flow mechanism of thixotropic substances. Industrial and Engineering Chemistry. 51: 856–857.
Hamann, D.D. 1983. Structural failure in solid foods. 532 p. *In*: Oeleg, M. and Bagley, E.B. (eds.). Physical Properties of Foods. AVI Publishing Company, Westport, Connecticut, USA.
Hernandez, E., Chen, C.S., Johnson, J. and Carter, R.D. 1995. Viscosity changes in orange juice after ultrafiltration and evaporation. Journal of Food Engineering. 25: 387–396.
Hertz, H. 1896. On the contact of elastic solids. pp. 146–162. *In*: Jones, D.E. and Schott, G.A. (eds.). Miscellaneous papers. English translation by Macmillan and Co. Ltd., London.

Holdsworth, S.D. 1971. Applicability of rheological models to the interpretation of flow and processing behavior of fluid food products. Journal of Texture Studies. 2: 393–418.
Holdsworth, S.D. 1993. Rheological models used for the prediction of flow properties of food products: a literature review. Transactions of the American Institute of Chemical Engineers. Part C. 71: 139–179.
ISO. 2008. Sensory analysis-Vocabulary. International Organization for Standards, ISO 5492-2008. P 107.
Ibarz, A., Gonzales C. and Explugas S. 1996a. Rheology of clarified passion fruit juices. Fruit Processing. 8: 330–333.
Ibarz, A., Garvin, A. and Costa, J. 1996b. Rheological behaviour of sloe (*Prunus spinosa*) fruit juices. Journal of Food Engineering. 27: 423–430.
Isikli, N. and Karababa, E. 2005. Rheological characterization of fenugreek paste (çemen). Journal of Food Engineering. 69: 185–190.
Juszczak, L. and Fortuna, T. 2004. Effect of temperature and soluble solid content on the viscosity of cherry juice concentrate. International Agrophysics. 18: 17–21.
Kaya, A. and Belibağli. K.B. 2002. Rheology of solid Gazıantep pekmez. Journal of Food Engineering. 54: 221–226.
Kokini, J.L. and Cussler, E.L. 1987. The psycho physics of fluid texture. *In*: Moskowitz, H.R. (ed.). Food Texture, Marcel Dekker, New York, USA.
Krokida, M.K., Maroulis, Z.B. and Saravacos, G.D. 2001. Rheological properties of fluid fruit and vegetable puree products: compilation of literature data. International Journal of Food Properties. 4: 179–200.
Lapasin, R. and Pricl, S. 1995. Rheology of Industrial Polysaccharides: Theory and Applications. Blackie Academic and Professional, New York, USA.
Liu, M., Haghighi, K. and Stroshine, R.L. 1989. Viscoelastic characterization of the soybean seedcoat. Transactions of the ASAE. 32: 946–952.
McCarthy, O.J. 1987. Large deformation testing of foods. Food Technology in New Zealand. July, pp. 40–43, Aug., pp. 14–20.
Mizrahi, S. 1979. A review of physicochemical approach to the analysis of the structural viscosity fluid fruit products. Journal of Texture Studies. 10: 67–82.
Mohsenin, N.N. 1986. Physical Properties of Plant and Animal Materials. 2nd revised and updated edition. Gordon and Breach Science Publishers, New York, USA.
Murase, H. and Merva, G.E. 1977. Static elastic modulus of tomato epidermis as affected by water potential. Transactions of the ASAE. 20: 594–597.
Ng, T.S.K. and McKinley, G.H. 2008. Power-law gels at finite strains. Journal of Rheology. 52(2): 417–449.
Nguyen, Q.D., Jensen, C.T.B. and Kristensen, P.G. 1998. Experimental and modeling studiesof the flow properties of maize and waxy maize starch pastes. Chemical Engineering Journal 70: 165–171.
Nijenhuis, K. 1997. Thermoreversible networks. Viscoelastic properties and structure of gels, Advances in Polymer Sciences. Springer, Berlin, Germany.
Nindo, C.I., Tang, J., Powers, J.R. and Singh, P. 2005. Viscosity of blueberry and raspberry juices for processing applications. Journal of Food Engineering. 65: 343–350.
Oomah, B.D., Sery, G., Godfrey, D.V. and Beveridge, T.H.J. 1999. Rheology of sea buckthorn (*Hippophae rhamnoides* L.) juice. Journal of Agricultural and Food Chemistry. 47: 3546–3550.
Peleg, M. 1980. Linearization of relaxation and creep curves of solid biological materials. Journal of Rheology. 24: 451–463.
Pitt, R.E. 1984. Stress-strain and failure characteristics of potato tissue under cyclic loading. Journal of Texture Studies. 11: 818–837.
Prins, A. and Bloksma, A.H. 1983. Guidelines for the measurement of rheological properties and the use of existing data. pp. 185–203. *In*: Jowitt, R., Escher, F., Hallstrom, B., Meffert, H.F.Th., Spiess, W.E.L. and Vos, G. (eds.). Physical Properties of Foods. Elsevier Applied Science. London.

Rao, M.A. 1977. Measurement of flow properties of fluid foods—developments, limitations and interpretation of phenomena. Journal of Texture Studies. 8: 257–282.
Rao, M.A. 1987. Predicting the flow properties of food suspensions of plant origin. Food Technology. 41(3): 85–88.
Rao, M.A. and Palomino, L.N.O. 1974. Flow properties of tropical fruit purees. Journal of Food Science. 39: 160–161.
Rao, M.A. 1986. Rheological properties of fluid foods. *In*: Rao M.A. and Rizvi, S.S.H. (eds.). Engineering Properties of Foods, Academic Press, San Diego, USA.
Rao, M.A., Cooley, H.J., Nogueira, J.N. and McLellan, M.R. 1986. Rheology of apple sauce: effect of apple cultivar, firmness, and processing parameters. Journal of Food Science. 51: 176–179.
Razavi, S.M.A. and Karazhiyan, H. 2012. Rheological and textural characteristics of date paste. International Journal of Food Properties. 15: 281–291.
Reiner, M. 1964. The Deborah Number, Physics Today. 17: 62.
Saiedirad, M.H., Tabatabaeefar, A., Borghei, A., Mirsalehi, M., Badii, F. and Varnamkhasti, M.G. 2008. Effects of moisture content, seed size, loading rate and seed orientation on force and energy required for fracturing cumin seed (*Cuminum cyminum* L.) under quasi-static loading. Journal of Food Engineering. 86: 565–572.
Saravacos, G.D. 1970. Effect of temperature on viscosity of fruit juices and purees. Journal of Food Science. 35: 122–125.
Sayyah, A.H.A. and Minaei, S. 2004. Behavior of wheat kernels under quasi-static loading and its relation to grain hardness. Journal of Agriculture Science and Technology. 6: 11–19.
Seely, F.B. and Smith, J.O. 1965. Advanced Mechanics of Materials. John Wiley & Sons, New York, U.S.A.
Shigley, J.E. and Mishke, C.R. 2001. Mechanical Engineering Design, 6th edn. McGraw-Hill, New York, U.S.A.
Steffe, J.F. 1996. Rheological Methods in Foods Process Engineering, 2nd edn., Freeman Press, East Lansing, Michigan, U.S.A.
Suárez-Quintanilla, D.C., Macedo-Ramírez R.C. and Vélez-Ruiz, J.F. 2003. Rheological behavior of three non-processed fruit fluids (papaya, melon and watermelon) and their concentrates. 3rd International Symposium on Food Rheology and Structure. 429–430.
Telis-Romero, J., Telis, V.R.N. and Yamashita, F. 1999. Frictional factors and rheological properties of orange juice. Journal of Food Engineering. 40: 101–106.
Timoshenko, S.P. and Goodier, J.N. 1970. Theory of Elasticity, 3rd edn. McGraw-Hill, New York, U.S.A.
Treloar, L.R.G. 1975. The Physics of Rubber Elasticity. Oxford University Press, Oxford.
Weltman, R.N. 1943. Breakdown of thixotropic structure as a function of time. Journal of Applied Physics. 14: 343–350.
Winter, H.H. and Chambon, F. 1986. Analysis of a crosslinked polymer at the gel point. Journal of Rheology. 30: 367–382.
Yu, C. and Gunasekaran, S. 2001. Correlation of dynamic and steady flow viscosities of food materials. Applied Rheology. 11: 134–140.
Zhong, Q. 2003. Cooling effects on the functionality and microstructure of processed cheese. Ph.D. Thesis, North Carolina State University, USA.

2

Pigments and Color of Muscle Foods

Jin-Yeon Jeong,[1] Gap-Don Kim,[2] Han-Sul Yang[1] and Seon-Tea Joo[1],*

ABSTRACT

Color is considered the most important sensory property, having a significant influence on the consumer's decision. Numerous studies have been conducted on the factors influencing the color of muscle foods. In this review, muscle pigment and myoglobin (Mb) chemistry are explored with the current and advanced methods to measure color of muscle foods. The color of muscle foods depends on concentration of heme pigments and their chemical state. Pigment concentrations in muscle foods are dependent on age, breed, sex, species, and muscle of the animal. Mb can exist in one of three forms: deoxyMb, oxyMb, or metMb in intact muscle. MetMb reducing enzyme systems allow the maintenance of reduced Mb or to reduce metMb, resulting in extending the shelf-life of muscle foods. The color of meat may be measured by several methods including reflectance, visual evaluation, instrumental color measurements, and computer vision analysis.

---

[1] Department of Animal Science, Institute of Agriculture & Life Science, Gyeongsang National University, 501 Jinjudaero, Jinju 660-701, South Korea.
[2] Department of Food Science & Biotechnology, Kyungnam University, 7 Kyungnamdaehak-ro, Changwon 631-701, South Korea.
* Corresponding author

2.1 Introduction

The overall appearance is the primary factor for consumers when purchasing muscle foods such as beef, pork, poultry, and fish. In particular, color is the most important sensory property, because it strongly influences the consumer's decision (Carpenter et al. 2001; Jeong et al. 2009). For example, consumers have a preference for a bright red color for fresh beef and have a definite bias against tan or brown discoloration. Moreover, the color of muscle foods is often used as an indicator of the wholesomeness of meat cuts, but acceptable meat color is short-lived during storage and display. Oxidation of pigments in muscle impacts meat color, and makes it less saleable because it is perceived as spoilt (Faustman and Cassens 1990). Discolored meat cuts are often minced and marketed in a reduced value form, because consumers are unlikely to accept muscle foods that are discolored.

There are many factors that influence the color of muscle-based foods, such as concentration of heme pigments and the chemical state of these pigments (Renerre and Bonhomme 1991; Hector et al. 1992). The color of meat is determined mainly by myoglobin (Mb) concentration in muscle, although hemoglobin (Hb) and cytochromes also have a minor contribution to meat color (Mancini and Hunt 2005). Total amount of heme pigments depends on muscle characteristics such as age, sex, breed, and species of the animal (Rickansrud and Henrickson 1967). In particular, Mb concentration is closely related to muscle fiber type composition (i.e., white fiber vs. red fiber) (Seideman et al. 1984). Meat color is strongly affected by the chemical state of Mb and by the MetMb reducing enzyme systems during storage (Jeong et al. 2009). The color of meat, i.e., Mb concentration and redox forms in meat, may be measured by several methods including reflectance, visual evaluation, instrumental color measurements, and computer vision analysis. In this chapter, muscle pigment and Mb chemistry will be explored with the current and advanced methods to measure color of muscle foods.

2.2 Pigments Concentration in Muscles

Pigment concentrations in muscle foods are primarily dependent on species, age, breed, sex, and muscle of the animal (Table 2.1). The color of meat is governed by the concentration of Mb with a minor contribution by Hb and the cytochromes. Most of the muscle Hb, the blood pigment, is removed during exsanguination at slaughter, but not completely. Thus, some muscle may retain 6–25% Hb of total heme pigments depending upon its anatomical location and the age, sex, and species of the animal (Rickansrud and Henrickson 1967). Cytochromes are present in muscle

Table 2.1 Concentration of myoglobin (mg/g) within muscles of different species.

Species	Muscle	Myoglobin Conc. (mg/g wet weight)	Reference
Cattle		(2–5)	Hunt and Hedrick (1977)
	Longissimus	3.48	Renerre and Labas (1984)
	Psoas major	3.71	
	Gluteus medius	4.11	
	Semimebranosus(outer)	3.91	
	Semimebranosus(inner)	3.56	
	Semitendinosus(outer)	2.97	
	Semitendinosus(inner)	1.95	
	Diaphragma	~7	
Pig (Hampshire)		(3–6)	Topel et al. (1966)
	Longissimus	2.94	
	Psoas major	6.37	
	Semimebranosus(outer)	4.05	
	Biceps femoris	5.06	
	Rectus femoris	5.66	
Chicken		(0.1–5)	Nishida and Nishida (1985)
	Pectoralis	~0.1	
	Vastuslateralis	2.8	
	Vastusintermedius	5.0	
	Biceps femoris	0.7	
	Rectus femoris	2.5	
	Gizzard	19	
Tuna		(0.5–20)	Brown (1962)
	Light meat	0.7	
	Lateral line muscle	20	
Swordfish	Red muscle	4.7–6.2	Reynafarje (1963)
	White muscle	0.9–2.11	Millikan (1939)
Horse mackerel	Red muscle	8.0–9.3	Millikan (1939)
	White muscle	0.36–0.40	
Bluefin tuna	Red muscle	15.1–23.7	Milikan (1939)
	White muscle	1.30–2.20	Lawrie (1953)
Human		4.4–5.2	Möller and Sylvén (1981)
Whale		60	Scholander (1940)
Dolphins		50–72	Dolar et al. (1999)
Seal		80	Robinson (1939)

foods at very low concentrations and, thus, have little effect on meat color, although cytochrome c in poultry meat constitutes 2–4% of total heme pigments (Pikul et al. 1986).

Significant differences occur in the color range among muscle foods from different species, and these differences are due to the different pigment contents in the muscle. Pigment contents in muscle are directly related to final meat color, because they increase color intensity from white or pink to red. For example, beef is redder and darker in color compared to lamb or pork due to higher pigment contents in bovine muscle. Lamb contains about one-half of this amount, and pork only one-fifth or less (Joo and Kim 2011). Whale muscle has a high Mb concentration and is darker in color followed by cattle, lamb, pig, and poultry in decreasing order of darkness (Moller and Skibsted 2006). Fish muscle varies considerably in pigment concentration, as some muscles are essentially white in appearance while others vary from white to red. The color of fish muscle is primarily related to the muscle fiber type composition in muscle and depends on the depth and diving habits of the fish, which dictate the need for greater oxygen storage (Kim and Hunt 2011).

As the animal ages, the pigment content increases in the muscles. The chronological age of an animal directly affects the darker color due to the increase in Mb concentration. As animals get older, the affinity of oxygen for Mb decreases; thus, they need to synthesize more Mb to store oxygen. Consequently, muscle from older or mature animals is darker than muscle from younger animals. Pigment accumulation with age is due to increased deposition in existing red muscle fibers. Veal contains only one-tenth the Mb as beef from a mature animal (Joo and Kim 2011).

Pigment concentration is also affected by breed and sex of the animal. Breed differences and genetic factors contribute to meat color by affecting Mb content and other inherent metabolic characteristics such as mitochondrial oxygen consumption and MetMb reducing capacity. In cattle, Mb concentration is highly heritable (0.85). Mb content is lowest in Charolais and Limousin followed by Angus, Red Angus, and Hereford (intermediate), and highest in Simmental and Gelbvieh carcasses (King et al. 2010). Furthermore, male animals are likely to have a greater Mb concentration and darker meat color than that of females. Castrated beef (steer) appears lighter in color than non-castrated bull beef (Seideman et al. 1984).

In contrast, muscles within a carcass vary in pigment content based on their physiological role. Mb contains heme iron, which is susceptible to oxidation and can lead to darkening, and muscle type and anatomical location influence the Mb concentration. For example, leg muscles are darker than loin muscles because their efficient work performance depends upon a higher proportion of red muscle fibers. There is an apparent color difference between thigh (red muscle) and breast (white muscle) in chicken due to different Mb concentrations (Lombardi-Boccia et al. 2002). The higher pigment content in red muscle fibers is mainly due to the need for Mb to store and deliver oxygen within the muscle. Thus, depending on muscle fiber type, Mb concentration can be different within the same muscle of the same animal (Hunt and Hedrick 1977).

Red muscle is redder and darker in color because it has more oxidative metabolic activity. Red muscle fibers (type I; slow twitch) contain a higher Mb concentration compared to that in white muscle fibers (type II; fast twitch), which has more glycolytic metabolism (Kim et al. 2010). In general, muscles used for locomotion appear redder and darker than muscles used for support. A locomotive muscle experiences more extensive muscle movement and requires a greater Mb concentration for oxidative metabolism than that of a support muscle (Meng et al. 1993). Additionally, Mb content in muscle is affected by factors such as exercise and diet of the animal as well as genetic and environmental factors (Livingston and Brown 1981).

2.3 Myoglobin Chemistry

2.3.1 Myoglobin and Derivatives

Meat color is strongly influenced by the chemical state of the Mb as well as Mb concentration. Mb is a metalloprotein composed of heme iron and globin, present in the sarcoplasmic fraction of the muscle, and its molecular weight is approximately 16,700 daltons. Mb consists of a single polypeptide protein or globin molecule, and an iron-containing heme prosthetic group (Fig. 2.1). The iron atom exists in various oxidation states, and the

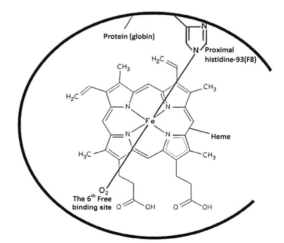

Figure 2.1 Myoglobin structure.

Color image of this figure appears in the color plate section at the end of the book.

porphyrin ring can be intact, oxidized, polymerized, or open. The heme iron is located in the center of the porphyrin ring and forms six bonds. It is ligated with four nitrogen atoms in the plane with the porphyrin ring. The other two coordination sites of these bonds connect perpendicular to this plane (Stryer 1995). The fifth site is occupied by the imidazole nitrogen of a histidine residue (His93) in the globin, whereas the sixth position is a free binding site that forms ligands with compounds such as oxygen, carbon monoxide, carbon dioxide, water, and nitrous oxide (Kim and Hunt 2011). As the sixth position of the iron remains available to bind electronegative atoms of various ligands, various binding formations at the sixth position are responsible for differences in meat color.

Globin is a single chain of 153 amino acids and a long protein containing a "pocket", which protects the heme iron group from oxidation (Clydesdale and Francis 1971). About 80% of amino acids that exist in α-helical conformations are arranged in eight domains wrapped around them to form a colorless, globular protein (Brewer 2004). The heme is located in a hydrophobic pocket lined with hydrocarbon portions of globin amino acids. The hydrophobic pocket provides the biological function of Mb. The non-polar environment inhibits the complete electron transfer between O_2 and Fe^{2+}, permitting the reversible complexing required for oxygen storage (Morrison and Boyd 1981). If the heme pocket is damaged by aminoacid modification or loss of some part of the peptide chain, it is less effective in maintaining the iron in the Fe^{2+} state, and Fe^{2+} is easily oxidized by O_2. Therefore, denaturation of the globin amino acids or peptides can increase oxidation of Mb, resulting in discoloration.

Mb can exist in one of three forms: deoxyMb, oxyMb, or metMb in intact muscle. These three Mb states interconvert, and the dominant Mb form depends on local conditions (Kropf 1993). DeoxyMb contains iron in the ferrous (Fe^{2+}) state and is characterized by the absence of a ligand at the sixth coordinate position of the heme group. It is frequently referred to as "myoglobin" or "reduced Mb", which is purplish-red in color. DeoxyMb is responsible for meat color immediately after cutting into deep muscle, or of vacuum packed meat. OxyMb is cherry-red in color and forms very quickly after exposure of deoxyMb to oxygen. OxyMb must be in the ferrous state for oxygenation to occur, and oxygen occupies the sixth binding site of the ferrous heme iron. Both deoxyMb and oxyMb readily oxidize to metMb, in which the heme iron has been oxidized to the ferric (Fe^{3+}) state and water occupies the sixth coordinate position. MetMb cannot bind oxygen and is, thus, physiologically inactive (Faustman and Cassens 1990). MetMb imparts the brown color that consumers associate with a lack of freshness and find unacceptable, whereas oxyMb gives meat a desirable color that consumers associate with freshness.

Several different configurations of Mb derivatives may form depending upon the ligand bound to the sixth coordination site of the heme ring (Fig. 2.2). Furthermore, exposing meat to different physical, chemical, and microbiological conditions substantially influences the formation of various Mb derivatives. When carbon monoxide (CO) binds to the free binding site of the native Mb, it forms a very bright color pigment called carboxyMb. CarboxyMb is more stable and lasts longer in the absence of oxygen; thus, exposure to aerobic conditions will slowly dissociate CO from carboxyMb. NitrosylMb may be generated during meat curing when deoxyMb reacts with nitrite. DeoxyMb provides nitrite reductase activity to form nitric oxide in biological systems (Hendgen-Cotta et al. 2008; Shiva et al. 2007). Because nitrosylMb is very similar to carboxyMb, it is quite stable and prolongs bright-red color for extensive periods under anaerobic conditions. However, once exposed to air, it rapidly forms an oxidized state called nitrosylMetMb. Upon heating, it is converted to nitrosylhemochromogen (also called nitrosohemochrome), which is responsible for the pinkish-red color of cured meats.

The chemical state of Mb is also affected by bacterial growth, which results in discoloration. SulfMb and choleglobin may be produced by reduced Mb reacting with bacterial by-products such as hydrogen sulfide

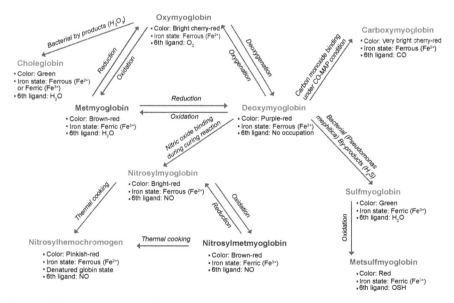

Figure 2.2 Color-affecting myoglobin redox interconversions under various oxidative, reducing, microbiological conditions and different chemical states of myoglobin.

Color image of this figure appears in the color plate section at the end of the book.

and hydrogen peroxide. Hydrogen sulfide production by *Pseudomonas mephitica* results in green discoloration on vacuum packaged meat (Nicol et al. 1970). SulfMb is generally found only in meat with high pH (> 6.0) and low oxygen tensions (1–2%) where bacteria can produce hydrogen sulfide (Kim and Hunt 2011). When meat is exposed to oxygen by opening a vacuum package, the green SulfMb forms the oxidized red color of metsulfMb, but it still smells of rotten egg due to hydrogen sulfide (Seideman et al. 1984). Bacterial growth can also produce hydrogen peroxide, which causes green discoloration of aerobically packaged meat by oxidizing Mb to choleglobin (Jensen and Turbain 1936).

2.3.2 Metmyoglobin Reduction

MetMb-reducing enzymatic systems are important for maintaining the fresh color of meat. These systems allow the maintenance of reduced Mb or to reduce MetMb, resulting in extending the shelf-life of muscle foods. The reduction process is primarily enzymatic in nature with nicotinamide adenine dinucleotide (NADH) as the coenzyme facilitating the conversion of ferric Mb to its ferrous form. NADH-cytochrome b5 MetMbreductase is the best-characterized enzyme involved in the reduction of oxidized Mb. The major components required for the enzymatic reduction of MetMb by that system are the enzyme (NADH-cytochrome b5 MetMbreductase), the intermediate (cytochrome b5), and the cofactor NADH (Bekhit and Faustman 2005).

The MetMbreductase enzyme has been purified and characterized from beef heart muscle (Hagler et al. 1979; Faustman et al. 1988), blue-white dolphin (Matsui et al. 1975), and bluefin tuna (Al Shaibani and Brown 1977). This enzyme is classified as a NADH-cytochrome b5 reductase, because it catalyzes the NADH-dependent reduction of cytochrome b5 (Livingston et al. 1985).The function of NADH-cytochrome b5 reductase is to transfer two electrons from NADH to two molecules of cytochrome b5 (Arihara et al. 1989). Reduced cytochrome b5 transfers the electrons to an acceptor MetMb, and NADH-cytochrome b5 reductase utilizes NADH to reduce ferricytochrome b5 to ferrocytochrome b5. Consequently, the ferrocytochrome b5 non-enzymatically reduces MetMb to ferrous Mb (Livingston et al. 1985).

Cytochromes are a group of intracellular hemoproteins that show a marked absorption spectrum in the visible region. Cytochrome b5 has three peaks in the absorption spectrum (a, b, and c) with maxima at 556, 526, and 423 nm, respectively, whereas the oxidized form of cytochrome b5 has one peak (c) at 413 nm (Hagihara et al. 1975). In general, cytochrome b5 is located in the endoplasmic reticulum where it can accept an electron from either NADH-cytochrome b5 reductase or nicotinamide adenine dinucleotide phosphate (NADPH)-cytochrome P450 reductase (Vergéres and Waskell 1995). Reduced cytochrome b5 provides reducing equivalents for the biosynthesis of selected lipids and some mixed function oxidases. The interaction between cytochrome b5 and NADH-cytochrome b5 reductase has been confirmed by direct enzyme titration (Lostanlen et al. 1980), nuclear magnetic resonance (Livingston et al. 1985), and the electrostatic potential of the surfaces of both proteins (Nishida and Miki 1996). Additionally, the involvement of this system in MetMb reduction in meat is evidenced by the presence of cytochrome b5 in the bovine muscle (Arihara et al. 1989).

The non-enzymatic reduction of MetMb occurs by NADH and NADPH in the presence of ethylenediaminetetraacetate (EDTA). NADH alone can reduce MetMb, particularly in the presence of EDTA, and the reduction increases considerably by mediation with flavins or methylene blue (Brown and Snyder 1969). Flavins catalyze the reduction of MetMb, and EDTA acts as a reductant. The rate of non-enzymatic reduction of muscle is similar to that of enzymatic reduction. NADH is the ultimate source of reducing equivalents for both reducing systems in porcine muscle and other animal species (Mikkelsen et al. 1999). However, detecting the non-enzymatic reduction of MetMb in bovine and ovine muscle has failed (Faustman et al. 1988; Hagler et al. 1979; Lanier et al. 1978; Reddy and Carpenter 1991; Renerre and Labas 1987). Despite some evidence, the NADH pool is unknown. It could involve mitochondria and/or sub-mitochondrial particles with the reversal of electron transport, which would allow reduction of NAD (Giddings 1974). NAD(H) decays more or less rapidly in muscles during storage (Faustman and Cassens 1990). The decay of NAD is due to complete loss of structural integrity and functional properties of the mitochondria (Renerre 1984; Ledward 1985). Consequently, a loss of reducing activity in meat during storage is related to a combination of factors including a fall in muscle pH and depletion of required substrates and co-factors. MetMb reductase (MRA) activity may be reduced by the increase of NADH activity in beef muscle, especially m. *psoas major* during cold storage (Jeong et al. 2009). For this reason, the freeze-thaw process accompanied by the damage to cellular compartments of muscle accelerates color deterioration of beef muscle (Jeong et al. 2011).

2.4 Measurement of Pigments and Meat Color

There are two main ways to measure the color of muscle foods; chemically, by analyzing the heme pigments in muscle and, physically, by measuring the light reflectance on the surface of meat. Chemical analysis methods are available for objectively measuring the color of muscle foods, and depend on the extraction of pigments (Hb and Mb) from the muscle followed by spectrophotometric determination of heme pigment concentrations (Hornsey

1956). However, these pigment extraction methods are time consuming and tedious. Consequently, several methods using light reflectance at the surface of muscle foods have been developed to objectively measure meat color.

2.4.1 Reflectance Measurements

The concentration of Mb and the percentages of Mb forms can be estimated with reflectance measurements of meat. Reflectance measurements are closely related to the amount and chemical state of Mb in meat *in situ*, and are affected by muscle structure, surface moisture, fat content, additives, and pigment concentrations (Fernández-López 2004). Reflectance values of different Mb redox forms may be equal at several wavelengths, i.e. isobestic points exist. Thus, deoxyMb, oxyMb and metMb can be quantified rapidly and easily using this procedure (Pérez-Alvarez and Fernández-López 2009).

It is well known that Mb absorbs light in the visible and ultraviolet regions (Fox 1966). Above 500 nm, OxyMb has absorption maxima at around 545 nm and 585 nm whereas below 500 nm, the maximum absorbance of metMb, oxyMb and deoxyMb are observed at 410, 418, 419, and 434 nm, respectively (Millar et al. 1996). In beef, several isobestic points are found at 474, 525, 572, and 610 nm, while all Mb redox forms are found at 525 nm (Synder 1965). This is similar in pork and chicken as well, while more isobestic wavelengths (430, 440, 450, 460, 510, 560, 570, 610, and 690 nm) are observed in chicken meat (Pérez-Alvarez and Fernández-López 2009). According to Krzywicki (1982), absorbance of the extracts at 572, 565, 545, and 524 nm is necessary to determine heme pigment concentrations in meat, and the total concentrations of the Mb redox forms may be calculated with these absorbance values. Additionally, reflectance differences at two wavelengths (630 minus 580 nm) or the ratio of 630/580 nm are used to determine the degree of discoloration due to MetMb formation.

Although several isobestic wavelengths are widely used to determine heme pigment concentrations, a method to quantitatively determine specific Mb redox forms is critical for diagnostics and trouble-shooting (Mancini and Hunt 2005). In general, equations for calculating Mb redox forms involve ratios of isobestic wavelengths specific to either 0% or 100% of the redox states. However, Mb values are occasionally calculated as either negative or > 100%, which are obviously inappropriate. To address these problems, the widely used equations of Krzywicki (1982) have been critically reevaluated by Tang et al. (2004), and new wavelengths have been suggested for determining metMb (545–503 nm), deoxyMb (565–557 nm), and oxyMb (572–582 nm).

2.4.2 Visual Evaluation

Color evaluations of muscle foods may be performed either by visual assessment or by instrumental color measurements. Sensory color evaluation data are critical to evaluate muscle foods, because they are directly related to consumer purchasing behavior. There is a strong relationship between color preference and purchasing intent with consumers discriminating against beef that is not bright red (Carpenter et al. 2001). Visual color assessment provides the degree of pigment, extent of oxygenated lean color, discoloration, and the amount of color change over time. However, the consistency and accuracy of a visual evaluation can be influenced by various conditions such as individual preference, lighting conditions, and visual deficiencies of the panelists (Kim and Hunt 2011).

The visual color of muscle foods may be determined by trained personnel or consumer panels. In general, descriptive color scales such as very bright red, bright red, and dull red are used by trained panelists, whereas consumer panels use hedonic scales such as acceptability, desirability, degree of like, or willingness to purchase. Panelist descriptions of meat color depend on individual cognition when references are not used. In this case, it is necessary for visual panels to be trained, screened, and selected based on their ability to consistently evaluate the desired color traits. Furthermore, visual panelists should have normal color vision and some ability to detect small differences in hue. The Farnsworth–Munsell 100 Hue Test can be used to screen and evaluate potential panelists (Kim and Hunt 2011). Pictorial color standards or color chips with built-in textural traits are very useful for training panelists and to help improve consistency.

Lighting type and intensity are very important to obtain reliable meat color data, because the lighting used for visual color evaluation can dramatically influence visual color perception. Therefore, lighting type and intensity must be standardized from sample to sample, panelist to panelist, and replication to replication. Fluorescent lighting (1614 lux, 150 foot candles) is recommended for a visual evaluation of meat, and it should have a color temperature of 3000–3500 K (lamps such as Deluxe Warm White, Natural, Deluxe Cool White, SP 3000, SP 3500). Note that fluorescent lighting has virtually no luminance in the red region. However, incandescent light likely increases beef and pork color desirability because its spectrum includes more red wavelengths (Barbut 2001). Lighting that is cool and white or lamps producing unrealistic pink, blue, or green tints should be avoided.

2.4.3 Instrumental Color Measurement

Color requires a light source that illuminates an object, and the observer senses the light that reflects or transmits the color of the object. Three quantitatively definable dimensions (lightness, chroma and hue) provide the stimulus that the brain converts into human perception of color (AMSA 1991). These properties of light reflected by the surface of muscle foods are correlated with panel scores (Hunt 1980). The results of reflectance measurements may be expressed using one or more of several color data systems such as Hunter Lab, CIE values (CIE XYZ), Munsell values (hue and chroma angles, and lightness values), or CIELAB (Pérez-Alvarez and Fernández-López 2009).

These instrumental color measurements may be used when human evaluations are not possible. Instrumental measurements provide excellent support data for visual panel evaluations, supplying objective information that helps define, diagnose, or explain changes in proportions of the chemical states of Mb. However, important criteria should be specified for an instrumental color methodology, because procedural details needed for a more universal use of data are often not mentioned (Tapp et al. 2011). Therefore, the criteria should be specified for instrumental details, including (1) types of instruments (colorimeters or spectrophotometers; Minolta, Hunter, Mini Scan, or others); (2) color system (Hunter, CIE, or tristimulus); (3) observer angle (2° or 10°); (4) illuminants (A, C, D65, or F); (5) instrument aperture sizes (0.64–3.2 cm); (6) procedure details such as number of readings and sample preparation for blooming time and temperature (Kim and Hunt 2011).

The most commonly used color space is the CIE L*a*b* with a 10° observer angle. L* is a measure of lightness in which 0 equals black and 100 equals white; positive values of a* indicate redness, whereas negative values indicate greenness; b* values indicate yellowness to blueness. Color may also be evaluated using parameters of hue angle $\tan^{-1}(b*/a*)$ and chroma $(a^{*2} + b^{*2})^{1/2}$. Hue angle is the development of color from red to yellow, and large angles indicate a less red product; thus, a higher value is associated with greater discoloration. Chroma indicates the saturation of color (i.e., color intensity or vividness), and a low value indicates a low color intensity where grey may more easily dominate.

When surface meat color is measured with the CIE L*a*b* color space, it is important to indicate the illuminant factor. Illuminant A (2854 K) is the recommended illuminant to use for meat products, because it places more emphasis on the red portion of the visible spectrum (Kim and Hunt 2011). It represents incandescent light and is richer in the red end of the visible spectrum. Different instrument aperture sizes can also influence color measures of meat. Smaller aperture sizes change the tristimulus values for any illuminant. In general, a* and b* values from most instruments are smaller when larger apertures are used (Yancey and Kropf 2008). Currently, Minolta-branded instruments (Tokyo, Japan) are used the majority of times to measure CIE L*a*b* on the surface of meats (Tapp et al. 2011).

Although CIE L*a*b* is commonly used to measure meat color, other color spaces are also useful for measuring color of specific muscle foods. Selecting the most appropriate color variable is project specific and dependent on the objectives of each experiment. For example, tristimulus coordinates (XYZ) are useful for measuring lamb carcass color (Alcalde and Negueruela 2001). Additionally, Hunter measurements for L, a, and b (illuminant A) are best for measuring the color of dry-cured ham (Garcia-Esteben et al. 2003). Hunter Lab and CIEL*a*b* provide more reproducible lightness data than that of L*u*v* and XYZ systems (Mancini and Hunt 2005).

In contrast, near-infrared (NIR) reflectance and visible reflectance by spectral analysis provides complete information about the molecular bonds and chemical constituents in a scanned sample. Thus, it can identify bands for deoxyMb (440 nm), metMb (475 nm), oxyMb (535 nm), and sulfMb (635 nm) (Liu et al. 2003). Furthermore, NIR spectra between 833 and 2035 nm may be used to predict L* and b* of beef loin and to predict beef steak color during ageing (Leroy et al. 2004). Non-invasive NIR tissue oximetry measures the amount of Mb redox forms both at and just below the surface. The most useful advantage of this methodology is that optical devices coupled to computers have the potential to offer very fast data acquisition from only a small surface area (Andrés 2007). This methodology also offers a new way to study pigment dynamics below the surface and the changes on the surface of the meat (Kim and Hunt 2011).

2.4.4 Computer Vision Analysis

Computer vision based on an analysis of digital camera images can be used to assess meat color. This method has distinct advantages over traditional color evaluation of muscle foods. Several benefits are associated with digital camera-derived jpeg images. Only a single digital observation is needed for a representative assessment of color compared with a colorimeter. Moreover, digital image data may account for surface variation in the Mb redox state and may be converted to numerous color measurement systems such as Hunter, CIE, and XYZ. After acquiring the digital image using a color camera and removing background, fat, and bone, L*a*b* values and RGB (red, green, blue) may be measured on a given area of the jpeg image (O'Sullivan et al. 2003).

Computer vision is a promising method for predicting the visual color of meat. It can predict sensory color scores with statistical and/or neural network models. For example, pork loin visual color may be predicted by computer vision, particularly digital image analysis combined with either a neural network or statistical modeling (Lu et al. 2000). The color scores predicted using a multi-layer neural network are significantly correlated with original trained panelist sensory scores. The neural network results in negligible prediction errors for 93% of the samples, while 84% of the samples have negligible errors when statistical models are used (Lu et al. 2000).

In addition, instrumental color measures derived from digital camera images can predict red and brown sensory terms (O' Sullivan et al. 2003). Compared to the colorimeter, which is based on point-to-point measurements, computer vision is more representative of sensory descriptors, because the camera measures the entire meat surface. Thus, the percentage of the surface covered by metMb can be estimated and the discolored areas on the surface of meat can be mapped using special software (Ringkob 2001). It is generally accepted that image analysis is useful for assessing pork color, particularly in detecting differences between yellow and white fat.

2.5 Conclusion

The color of muscle foods depends on heme pigments and chemical state of Mb. Total heme pigments are dependent on species, age, breed, sex, and muscle of the animal. There are basically three chemical states of Mb including deoxyMb, oxyMb, and metMb. Several different forms of Mb derivatives (carboxyMb, nitrosylMb, and sulfMb) may form depending upon the ligand bound to the sixth coordination site of the heme ring. MetMb reducing enzymatic systems are important for maintaining the fresh color of meat and these systems allow the maintenance of reduced Mb or to reduce metMb, resulting in extending the shelf-life of muscle foods. There are two main ways to measure the color of muscle foods; chemically, by analyzing the heme pigments in muscle and physically, by measuring the light reflectance on the surface of meat. The concentration of Mb and the percentages of various Mb forms may be estimated with reflectance measurements of meat. Color evaluations of muscle foods may be performed either by visual assessment or by instrumental color measurements. Color requires a light source that illuminates an object and the observer senses the light that is reflected or transmitted by the object in a color-dependent manner. Computer vision based on analysis of digital camera images may be used to assess meat color. This method has distinct advantages over traditional color evaluation of muscle foods.

References

Al-Shaibani, K.A., Price, R.J. and Brown, W.D. 1977. Purification of metmyoglobinreductase from bluefin tuna. Journal of Food Science. 42: 1013–1015.
Alcalde, M.J. and Negueruela, A.L. 2001. The influence of final conditions on meat colour in light lamb carcasses. Meat Science. 57(2): 117–123.
AMSA. 1991. Guidelines for meat color evaluation. In Proceedings of the 44th Annual Reciprocal Meat Conference (pp. 1–17) Manhattan, KS.
Arihara, K., Itoh, M. and Kondo, Y. 1989. Identification of bovine skeletal muscle metmyoglobinreductase as an NADH-cytochrome b5 reductase. Japanese Journal of Zootechnical Science. 60: 46–56.
Andrés, A., Murray, I., Navajas, E.A., Fisher, A.V., Lambe, N.R. and Bünger, L. 2007. Prediction of sensory characteristics of lamb meat samples by near infrared reflectance spectroscopy. Meat Science. 76: 509–516.
Barbut, S. 2001. Effect of illumination source on the appearance of fresh meat. Meat Science. 59(2): 187–191.
Bekhit, A.E.D., Simmons, N. and Faustman, C. 2005. Metmyoglobin reducing activity in freshmeat : A review. Meat Science. 71: 407–439.
Brewer, S. 2004. Irradiation effects on meat color—a review. Meat Science. 68: 1–17.
Brown, W.D. 1962. The concentration of myoglobin and hemoglobin in tuna flesh. Journal of Food Science. 27: 26–28.
Brown, W.D. and Snyder, H.E. 1969. Nonenzymatic reduction and oxidation of myoglobin and hemoglobin by nicotinamide adenine dinucleotides and flavins. Journal of Biological Chemistry. 244: 6702–6706.
Carpenter, C.E., Cornforth, D.P. and Whittier, D. 2001. Consumer preferences for beef color and packaging did not affect eating satisfaction. Meat Science. 57: 359–363.
Clydesdale, F.M. and Francis, F.J. 1971. The chemistry of meat color. Food Product Development. 581–184.
Dolar, M.L.L., Suarez, P., Poganis, P.J. and Kooyman, G.L. 1999. Myoglobin inpelagicscall cetaceans. Journal of Experimental Biology. 202: 227–236.
Faustman, C. and Cassens, R.G. 1990. The biochemical basis for discoloration in fresh meat: A review. Journal of Muscle Foods. 1: 217–243.
Faustman, C., Cassens, R.G. and Greaser, M.L. 1988. Reduction of metmyoglobin by extracts of bovine liver and cardicac muscle. Journal of Food Science. 53: 1065–1067.
Faustman, C. and Cassens, R.G. 1990. Influence of aerobic metmyoglobin reducing capacity on color stability of beef. Journal of Food Science. 55: 1278–1283.
Fernández-López, J., Sayas-Barberá, E., Pérez-Alvarez, J.A. and Aranda-Catalá, V. 2004. Effect of sodium chloride, sodium tripolyphosphate and pH on color properties of pork meat. Color Res. Appl. 29: 67–74.
Fox, J.B. 1966. The chemistry of meat pigments. Journal of Agricultural and Food Chemistry. 14(2): 207–210.
Garcia-Esteben, M., Ansorena, D., Gimeno, O. and Astiasaran, I. 2003. Optimization of instrumental colour analysis in dry-cured ham. Meat Science. 63(3): 287–292.
Giddings, G.G. 1974. Reduction of ferrimyoglobin in meat. CRC Critical Reviews in Food Technology. 5: 143–173.
Hagihara, B., Sato, N. and Yamanaka, T. 1975. Type b cytochromes. pp. 549–593. In: Boyer, P.D. (ed.). The enzymes (Vol. 11.). New York: Academic press.
Hagler, L., Coppers, R.I. Jr. and Herman, R.H. 1979. Metmyoglobinreductase. Identification and purification of a reduced NADH-dependent enzyme from bovine heart which reduces metmyoglobin. Journal of Biological Chemistry. 254(14): 6505–6514.
Hector, D.A., Brew-Graves, C., Hassen, N. and Ledward, D.A. 1992. Relationship between myosin denaturation and the color of low-voltage-electrically-stimulated beef. Meat Science. 31: 299–307.

Hendgen-Cotta, U.B., Merx, M.W., Shiva, S., Schmitz, J., Becher, S., Klare, J.P., Steinhoff, H.J.R., Goedecke, A., Schrader, J.R., Gladwin, M.T., Kelm, M. and Rassaf, T. 2008. Nitrite reductase activity of myoglobin regulates respiration and cellular viability in myocardial ischeemia-reperfusion injury. Proceedings of the National Academy of Sciences. 105(29): 10256–10261.

Hornsey, H.C. 1956. The colour of cooked cured pork. I. Estimation of the nitric oxide-haem pigments. Journal of the Science of Food and Agriculture. 7: 534–450.

Hunt, M.C. and Hedrick, H.B. 1977. Profile of fiber types and related properties of five bovine muscles. Journal of Food Science. 42(2): 513–517.

Jeong, J.Y., Kim, G.D., Yang, H.S. and Joo, S.T. 2011. Effect of freeze-thaw cycles on physicochemical properties and color stability of beef Semimembranosus muscle. Food Research International. 44: 3222–3228.

Jeong, J.Y., Hur, S.J., Yang, H.S., Moon, S.H., Hwang, Y.H., Park, G.B. and Joo, S.T. 2009. Discoloration characteristics of 3 major muscles from cattle during cold storage. Journal of Food Science. 74(1): C1–C5.

Jensen, L.B. and Turbain, W.M. 1936. A delicate test for blood pigment. Journal of Food Science. 1: 275–276.

Joo, S.T. and Kim, G.D. 2011. Meat quality traits and control technologies. Control of Meat Quality. Kerala: Research Signpost. Ch. 1.

Kim, G.D., Jeong, J.Y., Hur, S.J., Yang, H.S., Jeon, J.T. and Joo, S.T. 2010. The relation between meat color (CIE L* and a*), myoglobin content, and their influence on muscle fiber characteristics and pork quality. Korean Journal for Food Science of Animal Resources. 30(4): 626–633.

Kim, Y.H.B. and Hunt, M.C. 2011. Advance technology to improve meat color. Control of Meat quality. Kerala: Research Signpost. Ch. 2.

King, D.A., Shackelford, S.D., Kuehn, L.A., Kemp, C.M., Rodriguez, A.B., Thallman, R.M. and Wheeler, T.L. 2010.Contribution of genetic influences to animal-to-animal variation in myoglobin content and beef lean color stability. Journal of Animal Science. 88(3): 1160–1167.

Kropf, D.H. 1993. Colour stability. Meat focus international (June). 269–175.

Krzywicki, K. 1982. The determination of heme pigments in meat. Meat Science. 7: 29–36.

Lanier, T.C., Carpenter, J.A., Toledo, R.T. and Reagan, J.O. 1978. Metmyoglobin reduction in beef systems as affected by aerobic, anaerobic and carbon monoxide-containing environments. Journal of Food Science. 43: 1788–1992.

Lawrie, R.A. 1953. The relation of energy-rich phosphate in muscle to myoglobin and to cytochrome-oxidase activity. Biochem. J. 55: 305–309.

Ledward, D.A. and Shorthose, W.R. 1971. A note on the haem pigment concentration of lamb as influenced by age and sex. Animal Product. 13: 193–195.

Ledward, D.A. 1985. Post-slaughter influences on the formation of metmyoglobin in beef muscles. Meat Science. 15: 149–171.

Leroy, B., Lambotte, S., Dotreppe, O., Lecocq, H., Istasse, L. and Clinquart, A. 2004. Prediction of technological and organoleptic properties of beef Longissimus thoracis from near-infrared reflectance and transmission spectra. Meat Science. 66: 45–54.

Livingston, D.J. and Brown, W.D. 1981. The chemistry of myoglobin and its reactions. Food Technology. 35(2): 238–252.

Livingston, D.J., McLchlan, S.J. Lamar, G.N. and Brown, W.D. 1985. Myoglobin: cytochrome b5 interactions and the kinetic mechanism of metmyoglobinreductase. Journal of Biological Chemistry. 260: 15699–15707.

Liu, Y., Lyon, B.G., Windham, W.R., Realini, C.E., Pringle, T.D. and Duckett, S. 2003. Prediction of color, texture and sensory characteristics of beef steaks by visible and near infrared reflectance spectroscopy. A feasibility study. Meat Science. 65(3): 1107–1115.

Lombardi-Boccia, G., Martínez-Domínguezb, B., Aguzzia, A. and Rincó n-León b, F. 2002. Optimization of heme iron analysis in raw and cooked red meat. Food Chemistry. 78: 505–510.

Lostanlen, D., Gacon, G. and Kapan, J.C. 1980. Direct enzyme titration curve of NADH: cytochrome b5 reductase by combined isoelectric focusing/electrophoresis. Interactions between enzyme and cytochrome b5. European Journal of Biochemistry. 112: 179–183.

Lu, J., Tan, J., Shatadal, P. and Gerrard, D.E. 2000. Evaluation of pork color by using computer vision. Meat Sceince. 56(1): 57–60.

Matsui, T., Shimizu, C. and Matsuura, F. 1975. Studies on metmyoglobin reducing systems in the muscle of blue white-dolphin.II. Purification and some physio-chemical properties of ferrimyoglobinreductase. Bulletin of Japanese Society of Scientific Fisheries. 41: 771–782.

Mancini, R.A. and Hunt, M.C. 2005. Current research in meat color. Meat Science. 71: 100–121.

Meng, H., Bentley, T.B. and Pittman, R.N. 1993. Myoglobin content of hamster skeletal muscle. Journal of Applied Physiology. 74(5): 2194–2197.

Millikan, G.A. 1939. Muscle hemoglobin. Physiol. Rev. 19: 503–523.

Möller, P. and Sylvén. C. 1981. Myogloin in human skeletal muscle. Scand. J. Clin. Lab Invest. 41: 479–482.

Moller, J.K.S. and Skibsted, L.H.O.L. 2006. Myoglobin—the link between discoloration and lipid oxidation in muscle and meat. Química Nova. 29(6): 1270–1278.

Millar, S.J., Moss, B.W. and Stevenson, M.H. 1996. Some observations on the absorption spectra of various myoglobin derivatives found in meat. Meat Science. 42(3): 277–288.

Mikkelsen, A., Juncher, D. and Skibsted, L.H. 1999. Metmyoglobin reductase activity in porcine m. longissimusdorsi muscle. Meat Science. 51: 155–161.

Morrison, R.T. and Boyd, R.N. 1981. Amino acids and protein. pp. 1132–1162. *In*: Organic Chemistry (3rd edn.).Boston, MA; Allyn and Bacon, Inc.

Nicol, D.J., Shaw, M.K. and Ledward, D.A. 1970. Hydrogen sulfide production by bacteria and sulfmyoglobin formation in prepacked chilled beef. Applied Environ. Microbiol. 19(6): 937–939.

Nishida, J. and Nishida, T. 1985. Relationship between the concentration of myoglobin and parvalbumin in various types of muscle tissues form chickens. Br. Poult.Sci. 26: 105–115.

Nishida, H. and Miki, K. 1996. Electrostatic properties deduced from refined structures of NADH-cytochrome b5 reductase and the other flavin-dependent reductases: pyridine nucleotide-dining and interaction with an electron transfer partner. Proteins. 26: 32–41.

O'Sullivan, M.G., Byrne, D.V. and Martens, M. 2003. Evaluation of pork colour: sensory colour assessment using trained and untrained sensory panelists. Meat Science. 63(1): 119–129.

Pérez-Alvarez, J.A. and Fernández-López, J.F. 2009. Color measurements on muscle-based foods. Handbook of Muscle Foods Analysis. CRC press: Tayler& Francis group. Ch. 26.

Pikul, J., Niewiarowicz, A. and Kupijaj, H. 1986. The cytochrome c content various poultry meats. Journal of the Science of Food and Agriculture. 37(12): 1236–1240.

Renerre, I.M. and Bonhomme, J. 1991. Effects of electrical stimulation, boning temperature and conditioning mode on display color of beef meat. Meat Science. 29: 191–202.

Reddy, I.M. and Carpenter, C.E. 1991. Determination of metmyoglobin reductase activity in bovine skeletal muscle. Journal of Food Science. 56: 1161–1164.

Renerre, M. and Lebas, R. 1987.Biochemical factors influencing metmyoglobin formation in beef muscles. Meat Science. 19: 151–165.

Renerre, M. and Labas, R. 1984. Variability between muscles and between animals of the color stability of beef meats. Sciences des Aliments. 4: 567–584.

Reynafarje, B.J. 1963. Simplifec method for the determination of myoglobin. J. Lab. Clin. Med. 61: 138–145.

Rickansrud, D.A. and Hendrickson, R.L. 1967. Total pigments and myoglobin content in four bovine muscles. Journal of Food Science. 32: 57–61.

Ringkob, T.P. 2001. Image analysis to quantify color deterioration on fresh retail beef. *In*: Proceeding 54th Reciprocal Meat Conference, 24–28 July 2001, Indianapolis, Indiana.

Robinson, D. 1939. The muscle hemoglobin of seals as an oxygen store in diving. Science. 90: 276–277.

Scholander, P.F. 1940. Experimental investigations on the respiratory function in diving mammals and birds. HvalradetsSkrift. 22: 1–131.

Seideman, S.C., Cross, H.R., Smith, G.C. and Durland, P.R. 1984. Factors associated with fresh meat color: A review. Journal of Food Quality. 6(3): 211–237.

Shiva, S., Huang, A., Grubina, R., Sun, J., Ringwood, L.A., MacArthur, P.H., Xu, X., Murphy, E., Darley-Usmar, V.M. and Gladwin, M.T. 2007.Dexoymyoglobin is a nitrite reductaseat generates nitric oxide and regulates mitochondrial respiration. Circ. Res. 100(5): 654–661.

Synder, H.E. 1965. Analysis of pigments at the surface of fresh beef with reflectance spectrophotometry. Journal of Food Science. 30: 457–459.

Stryer, L. 1995. Biochemistry. New York; W.H. Freeman and Company.

Tang, J., Faustman, C. and Hoagland, T.Q. 2004. Krzywicki revisted: Equations for spectrophotometric determination of myoglobin redox forms in aqueous meat extracts. Journal of Food Science. 69(9): 717–720.

Tapp, W.N., Yancey, J.W.S. and Apple, J.K. 2011. How is the instrumental color of meat measured? Meat Science. 89(1): 1–5.

Topel, D.G., Merkel, R.A., Mackintosh, D.L. and Hall, J.L. 1966. Variation of some physical and biochemical properties within and among selected porcine muscles. Journal of Animal Science. 25: 277–282.

Vergères, G. and Waskell, L. 1995. Cytochrome b5, its functions, structure and membrane topology. Biochimie. 77: 604–620.

Yancey, J.W.S. and Kropf, D.H. 2008. Instrumental reflectance values of fresh pork are dependent on aperture size. Meat Science. 79(4): 734–739.

3

Methodologies to Analyze and Quantify Lipids in Fruit and Vegetable Matrices

Hajer Trabelsi* and *Sadok Boukhchina*

ABSTRACT

A wide variety of methodologies have been developed and implemented to categorize vegetable oils according to their composition and characteristics. The most appropriate method must be carefully selected to achieve the goals, taking into account the nature of the analyte molecules and the experimental conditions. The chapter discusses the selection of the most appropriate methods for the analysis and quantification of different lipid components in vegetable matrices, such as fruit, seeds and leaves.

3.1 Introduction

Natural fats such as vegetable oils and animal fats are complex mixtures of different chemical compounds with the major gravimetric part being triglycerides, and the remainder made up by free fatty acids, partial glycerides, phospholipids, glycolipids and other unsaponifiable matter (Guichard 1967; Naudet 1992). It is this complexity, both in the nature and relative quantities of the various constituents, that gives the fats their

---

University of Tunis El Manar, Faculty of Sciences of Tunis, Unit of Biochemistry of Lipids and Proteins, 2092 El Manar II, Tunisia.
* Corresponding author

uniqueness, quality and identity. Faced with the growing issue of the adulteration of oil, one of the main interests in the vegetable oil sector is the search for the specific chemical make-up of oils that can be used to ensure their authenticity. This requires a good knowledge of the qualitative composition of the major and minor constituents of vegetable oils. Thus, to ensure a better determination of the quality of food oils, researchers continue to develop methods of extraction, separation and analysis (Tranchida et al. 2013; Korostynska et al. 2013).

The purpose of this chapter is to present techniques and methods used for qualitative and quantitative determination of different fractions of oils in fruit and vegetable matrices.

3.2 Methods for Vegetable Oil Extraction

The choice of lipid extraction method depends on the molecules or the lipid fraction which is sought for analysis. The Soxhlet method is the most commonly used for lipid extraction because it has the main advantage of extracting the maximum lipid content from vegetable samples by passing a nonpolar solvent through the sample for approximately 6 hours in repetitive cycles. Furthermore, this method requires much less solvent than the method consisting of successive macerations for the same extraction efficiency. However, a major limitation of the Soxhlet method is the application to other polar molecular species present in oil that might be important as chemical markers to the oil identity. Indeed, phospholipids are polar, whereas lipid reserves, mainly in the form of triglycerides, are neutral. Phospholipids and triglycerides both consist in part of fatty acids. For measurements of fatty acids total content (whether included as part of a polar or nonpolar molecules), it is necessary to extract lipids using a non-alternative method. Therfore a method which combines a polar solvent (to extract the phospholipids, in particular) and a nonpolar solvent (to extract the triglycerides) is necessary. The method of Bligh and Dyer (1959), derived from modifications to the method of Folch et al. (1957) is preferred for the determination of the molecular species of phospholipids and consists of grinding the lightly boiled (to denature the phospholipase enzyme) plant material in a chloroform-methanol mixture (2:1 v/v), followed by centrifugation of the mixture and recovery of the organic phase containing lipids. However, the drawback of this method is that it uses toxic solvents, and therefore is much less used in the determination of the total lipid content.

Methods involving heating have the additional inconvenience that the extraction process can lead to degradation of some chemicals.

A new method of oil extraction that has shown effectiveness in comparison to the conventional method was described by Bligh and Dyer

(1959). This method uses solvent–free, ultrasound-assisted extraction suitable for recovering oil from microalgae in reduced time and with a simple and scalable pre-industrial device (Fanny et al. 2012). The constraint of this method is that the amount of fatty acid methyl esters obtained seems to be low (0.2% by UAE method/5.7% by Bligh and Dyer method), however, their quality does not seem affected. The three previously mentioned methods are commonly used for the extraction of vegetable oils. Nevertheless, there are other known methods which are designed for more specific analytical goals.

3.3 Thin-layer Chromatography in Lipid Analysis

Even with relatively straight-forward mechanical simplicity, thin-layer chromatography (TLC) is still an essential method used for food composition analysis, including lipid analysis. Commonly in TLC the analyte solution is applied to a specific point on a thin silica plate (stationary phase), which is then positioned vertically in a tank of eluent (liquid phase). As the eluent rises over the stationary phase by capillary action, compounds are separated by their differential affinity towards the stationary phase. Once separated, lipids may be visualized through the application of developers such as iodine vapors (all lipids and mainly unsaturated lipids will then appear as dark spots) and $MnCl_2$ and methanol in a sulfuric acid bath, which shows different colors for each lipid. Another developer uses 2',7'-dichlorofluorescein which adheres to unsaponifiable matter and glows under UV light.

In a specific example, the unsaponifiable matter in *Pistacia lentiscus* fruit was separated into subfractions on preparative silica gel thin-layer plates (silica gel 60 G F254), using 1-dimensional TLC with hexane-Et_2O (65:35 by volume) as the developing solvent. After development the plate was sprayed with 2',7'-dichlorofluorescein and viewed under UV light. Several bands were observed, and through the use of standards, were identified as sterols, aliphatic alcohols, triterpenic alcohols and hydrocarbons. The bands were scraped off separately and extracted with $CHCl_3$–Et_2O (1:1), filtered to remove the residual silica, dried in a rotary evaporator and stored at 10 °C for GC-MS analysis (Herchi et al. 2009a; Trabelsi et al. 2012).

Papaya seed oil has been analyzed by silica gel preparative liquid chromatography (PLC) and analytical argentation TLC for the analysis of lipid classes including fatty acids and triglycerides. Application of bromine and sulfuryl chloride vapors allowed for quantification using densitometry at 450 nm (Nguyen and Tarndjiiska 1995). The fatty acid methyl ester content and their antioxidant activity in paprika were studied using silica gel TLC with acetone-light petroleum (15:85) mobile phase (Matsufuji et al. 1998). Using silica gel with chloroform-methanol-acetic acid-methanol-water

(35: 25: 4: 14: 2.2) and hexane-diethyl ether (4:1) mobile phases, soybean phospholipids were analyzed by double development detection with PMA and Dittmer reagents followed by densitometric quantification at 654 nm (Guo and Xu 1998). High-performance thin-layer chromatography (HPTLC) is often considered to be of traditional interest only because more sophisticated and more modern methods are available. But, according to Fuchs et al. (2011), HPTLC is one of the most versatile and reliable techniques for lipid analysis. It is an extremely powerful tool and can be applied to all relevant lipid classes. HPTLC has shown itself to be not only a fast, economical method but also one that may be used with minimum trouble-shooting. Finally, rather than risking damage to an HPLC column, HPTLC may be first applied to samples containing unknown components (for instance from food) (Fuchs et al. 2011).

For the analysis of all animal and plant lecithins and phospholipid mixtures, a standard German (DGF) HPTLC method has been highlighted (Lange and Fiebig 1999).

Recently, a new coupled method was developed that conveniently combines matrix-assisted laser desorption/ionization imaging mass spectrometry (MALDI-IMS) with TLC-blotting (TLC-Blot-MALDI-IMS). The combination of MALDI-IMS and TLC blotting enabled detailed and sensitive analyses of phospholipids such as phosphatidylethanolamine, phosphatidylinositol, phosphatidylserine, phosphatidylcholine and sphingomyelin (Zaima et al. 2011). This study showed the power of combining MALDI-IMS with HPTLC and it is evident that additional experimental considerations that lead to absolute quantification would have a huge potential.

Further improvements could be achieved by the introduction of Ultra-Thin-Layer Chromatography (UTLC) that enables even higher-precision separations (Fuchs et al. 2011).

3.4 Gas Chromatography in Lipid Analysis

Gas chromatography (GC) is a technique that allows separation and analysis of compounds in the gas phase. An inert gas mobile phase, such as helium or nitrogen, moves volatilized compounds through a column, which separate differentially based on interactions with the stationary phase. Naturally, the requirement of GC is that the analytes be sufficiently volatile and stable at the experimental temperatures may become a limiting factor in terms of compounds that can be analyzed without chemical modification. For example, fatty acids can be chemically modified with a volatilizing agent, such as tri-methylsilyl, increasing volatility and allowing them to be analyzed by GC.

66 Methods in Food Analysis

Many detectors using different principles have been developed for use with gas chromatography, however, only a few continue to be widely used. The flame ionisation detector (FID) has been used with a large collection of organic compounds, and offers high sensitivity, good stability and a linear response over a larger range (Christie 1989). GC-FID was used for fatty acid methyl esters analysis in *Pistacia lentiscus* fruit (Fig. 3.1) using a Innowax capillary column with a length of 30 m, internal diameter of 0.32 mm and film thickness 0.25 µm (Trabelsi et al. 2012).

The fatty acid methyl esters were identified by comparing sample retention times with those of standards, however, using the above column properties, it was not possible to efficiently separate longer isomeric species, nor fatty acid methyl esters with longer saturated chains. GLC analysis was also shown to be capable of detecting olive oil adulteration with other plant oils, through analysis of sterols using a polar column (Al-Ismail et al. 2010).

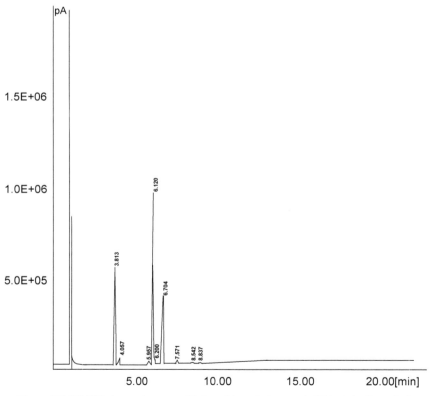

Figure 3.1 GC-FID chromatogram of fatty acid methyl esters in *Pistacia lentiscus* fruit.

Similarly, mass spectrometry (MS) is widely used today and offers a very good ability to detect, quantify and characterize lipids separated by GC. Structural eludication of α- and ß-regio-isomers of monopalmitoylglycerol (MAG C16:0) in their silylated forms have been shown to be possible with gas chromatography-mass spectrometry (GC-MS) with electron impact ionization (EI) (Destaillats et al. 2010). Although the relative amounts of ion fragmentation that occurs for a given chemical species may be influenced by the specific geometry of the mass spectrometer, data libraries containing the mass spectra of a vast number of compounds may be used to identify compounds in an unknown sample. The 4-desmethylsterols (Fig. 3.2) in linseed oil were identified using a GC-MS method by comparing their retention times and mass spectra with those of their pure molecules and those found on the NIST Mass Spectral Library (Herchi et al. 2009b; Trabelsi et al. 2012).

Recently, a new investigation has led to the development of a comprehensive two-dimensional GC (GC×GC) method, with dual MS/FID detection, for the qualitative and quantitative analysis of the entire unsaponifiable fraction of vegetable oils (Tranchida et al. 2013).

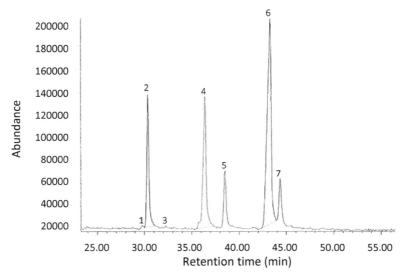

Figure 3.2 GC-MS chromatogram of 4-desmethylsterols: (1) cholesterol, (2) a-cholestanol(SI), (3) brassicasterol, (4) campesterol, (5) stigmasterol, (6) b-sitosterol, (7)D5-avenasterol (Herchi et al. 2009b).

3.5 High Performance Liquid Chromatography (HPLC) in Lipid Analysis

High-performance liquid chromatography (HPLC) is a highly universal type of analytical procedure with a wide variety of applications. Compared with GC, it offers the added benefit of separating non-volatile, high-molecular-mass compounds. Through the use of different stationary phases and mobile phase gradients, procedures can be greatly optimized for a compound class of interest, or for a sample which contains a complement of very dissimilar compounds. Several detection methods may be used in combination with HPLC, with ultraviolet-visible detector (UV-VIS) being the most commonly used. Other detection methods, such as refractive index (RI), FID, MS, evaporative light scattering (ELSD), fluorescence (FD) and electrochemical detection are also used (Cert et al. 2000).

Consequently, HPLC based techniques have found a wide use in lipid analysis. A normal phase HPLC equipped with a quaternary pump and an evaporative light scattering detector (ELSD) was used to quantify the individual phospholipids in *Jatropha curcus* seed. Identification of phospholipids was carried out by comparing the retention times of commercial standards to the signals arising from the phospholipids present in the sample (Rao et al. 2009). Calibration curves for each of the phospholipids were obtained by injecting different concentrations, and next were used to quantify the individual phospholipids following the method described by Avalli and Contarini (2005). Furthermore, the contents, classes and different molecular species of phospholipids in *Pistacia lentiscus* fruit oil were determined by LC-ESI-QTOF-MS, and MS/MS (Trabelsi et al. 2013). This technique has emerged as the most suited for routine, high-throughput analyses and is very effective for the analysis of phospholipids (Pulfer and Murphy 2003).

The separation of mono-, di- and triacylglycerols in olive oil, sesame seed oil and vegetable oil may be readily accomplished by HPLC with silver(I)-mercaptopropyl modified silica gel, or silver thiolate chromatographic material (AgTCM), which may be used with evaporative light scattering detection (ELSD) or atmospheric pressure chemical ionization mass spectrometry (APCI-MS). According to Dillon et al. (2012), the separation of triacylglycerols varying by degrees of unsaturation and *cis/trans* configuration in common oil samples may be achieved using a simple linear gradient of hexane and acetone. In addition to double bonds, AgTCM also displays major selectivity for compounds with different levels of polarity, allowing for efficient separation between mono-, di- and triglycerol classes. When coupled to a mass spectrometer, commonly using nano- or electrospray ionization, the identification process can be greatly improved. The composition of plant membrane lipids in monogalactosyldiacylglycerols (MGDGs) and digalactosyldiacylglycerols (DGDGs) may be investigated by reversed-phase high performance liquid chromatography/mass spectrometry (RP-HPLC-MS) with accurate mass measurement. Galactolipids were ionized by electrospray operated in the positive-ion mode and identified by their MS/MS spectrum (Zábranská et al. 2012). The tocopherols and tocotrienol compositions in hazelnut oil may be characterized by the use of the normal phase-high performance chromatography/mass spectrometry (NP-HPLC/MS) as the analysis technique (Benitez-Sanchez et al. 2003).

3.6 Mass Spectrometric Based Methods for Vegetable Oil Analysis

Lipids involve a broad range of naturally occuring chemical compounds. As previously referred, mass spectrometry (MS) has proven to be one of the most appropriate techniques for their characterization. Aside from offering high sensitivity and quantitation, its ability for chemical characterization is profound. In its simplest form, mass spectrometry measures ionized compounds mass to charge ratio (m/z). MS requires a system for the sample introduction, an ionization source to ionize the molecules of the sample, a mass analyzer separating the ions according to their m/z and a detector to count these ions. Electron impact ionization (EI) is the oldest but the most robust method of sample ionization. A limitation of this technique is that it requires compounds to be volatile and larger compounds tend to dissociate completely during the ionization process, making assignment of the m/z of the molecular ion difficult. A major breakthrough for the use of MS for lipid analysis (and in many other important fields) was the introduction of soft ionization techniques which allowed for the desoprtion and ionization of larger compounds without causing fragmentation. The most important for lipid analysis were the developments of electrospray ionization (ESI) and atmospheric pressure chemical ionization (APCI). MALDI (matrix assisted laser desorption ionization), when coupled to a TOF (time-of-flight) mass spectrometer, has also seen use in lipid analysis.

According to Horn et al. (2011), nanospray ESI/mass spectrometry is a method that combines the visualization of individual cytosolic lipid droplets (LDs), microphase extraction of lipid components from droplets, and direct identification of lipid composition, even to the level of a single LD. The molecular species of four derivatives of the sterol lipid class from *Arabidopsis thaliana* were identified by the use of a nanospray ionization quadrupole-time-of-flight mass spectrometry (Q-TOF/MS). Quantification of molecular species was achieved in the positive mode after fragmentation in the presence of internal standards and was validated by comparison with results obtained with TLC/GC (Wewer 2011). According to the author, the Q-TOF method is far more sensitive than GC or HPLC. Therefore, Q-TOF MS/MS provides a comprehensive strategy for sterol lipid quantification that can be adapted to other tandem mass spectrometers.

A major chemical difference between different lipid compounds is the total number of double bonds. Lipid species with different degrees of saturation can be detected easily with MS as the *m/z* is different, however, quantification is complex because there will be some isotopic overlap. This can be solved through a separation step prior to ionization, such as HPLC. The TAG composition of *Capparis spinosa* was determined using a direct injection electrospray ionization quadrupole-time-of-flight mass spectrometer (ESI-Q-TOF/MS) operating in positive ion mode. As the natural abundances of the elements are known, they may be used to estimate the extent of isotopic overlap, and the experimental mass spectrum may be deconvoluted to a corrected mass spectrum. Quantification of the TAG species may be obtained by comparing the intensities of endogenous TAG species with a TAG internal standard, such as tripalmitin, when it is absent from the sample (Tlili et al. 2011). TAGs profiles of amazonian vegetable oils and fats were determined using MALDI-TOF mass spectrometry fingerprinting. The oil was characterized without pre-separation or derivatization via dry (solvent-free) sample deposition. Mass spectra were acquired in the positive ion mode and desorption/ionization was accomplished using a UV laser (337 nm) (Saraiva et al. 2009). The quantitative aspects of MALDI MS are often considered to be lacking, particularly if complex lipid mixtures are to be analyzed. Although the detailed role of the matrix is not yet completely understood, it has been evident that careful choice of the matrix is important in order for all compounds of interest to be detected. Imaging MS is nowadays widely established and significant interest is paid in this context to the analysis of lipids that ionize particularly well and are, thus, more sensitively detectable in tissue slices than other biomolecules such as proteins (Fuchs et al. 2010).

A second important feature of most modern mass spectrometers is the ability to do tandem MS (MS/MS) experiments, where a certain *m/z* ion is mass-selected and made to undergo fragmentation, commonly with collision induced dissociation (CID). For example, phospholipids (PL) contain two fatty acid chains, and the total molecular *m/z* alone is not sufficient in identifying the exact fatty acids (FA) that make up the molecule, as multiple combinations could results in the same total *m/z*. CID of phospholipids ionized in the negative mode producing [PL-H]$^-$ ions results in cleavage of the fatty acid chains, yielding spectral peaks of [FA-H]$^-$ ions, which are used to identify the fatty acids present in the molecule. MS/MS is invaluable to the analysis of many other lipid compounds, including triglycerides, which dissociate in the positive mode in a manner allowing to identify the three FA chains.

3.7 Raman Spectroscopy for Vegetable Lipid Analysis

Spectroscopic methods are also widely used for analysis of lipids because of their many benefits, which include speed, relative cost-effectiveness, and usually simplified sample preparation (Baeten et al. 2001). Raman spectroscopy is a promising tool for the characterization and quantification of different properties of lipids (Baeten 2010), useful for determining total unsaturation (iodine value) of a sample, *cis/trans* geometrical isomer ratio, unsaturation of oil or fat (monounsaturated vs. polyunsaturated fatty acids), quantification of a particular triacylglycerol or fatty acid and crystal structure of specific fat products (Beattie et al. 2004; Muik et al. 2005). In fact, according to Baeten (1998), FT-Raman allows the analysis of several vegetable oils and provides information not only on the degree of unsaturation of oils, but also on their saturated fatty acid, monounsaturated fatty acid, and polyunsaturated fatty acid contents, and the results showed a high correlation with the respective fatty acid profiles obtained by gas chromatography.

Furthermore, it has been possible to calibrate a Raman spectrometer for the quantification of trilinolein, which is used as an indicator of adulteration in virgin olive oils (Baeten and Aparicio 2000). Raman spectroscopy is also a valid technique for analyzing minor compounds of oils and fats such as carotenoids in oils, or waxes or squalene in the unsaponifiable fraction of virgin olive oil (Baeten et al. 2005).

3.8 Nuclear Magnetic Resonance (NMR)

Nuclear magnetic resonance is based on the fact that the nuclei of many atoms have a spin, which causes them to behave as tiny magnets. Thus they are affected by any applied magnetic field. If, in addition to this magnetic field, there is an oscillating field in the radio frequency range, the nucleus will resonate between different energy levels at a definite frequency (Hopkins 1961).

One of the first applications of NMR was as a means to determine the authenticity of vegetable oils by the determination of fatty acid composition (Knothe 2004). Furthermore, during the methanolysis of triglycerides (TGs), the intermediates sn-1,2-, 1,3-diglycerides (DGs) and 1-, 2-monoglycerides (MGs) were identified by NMR. Thus, the results showed that a significant intermediate was sn-1,3-DGs in the DGs, but sn-2-MGs was not found, suggesting that the methanolysis reaction may occur easily at sn-2-

position for both TGs and DGs (Jin et al. 2007). However, with better suited techniques being applied to analyze adulterated oils, quantitative analysis becomes more important, and in the case of NMR, appears in the form of the intensity of a specific peak, which is characteristic for a specific oil (Hidalgo and Zamora 2003).

3.9 Capillary Electrophoresis

Capillary electrophoresis is an efficient method of analysis, based on the separation of charged species, under the effect of an electric field in a capillary tube with 50 to 100 microns in diameter, filled with an electrolyte solution. This technique has shown its value in analyzing complex mixtures of fatty acids from C_{10} to C_{24}, with normal and substituted chains (Mofaddel and Desbène-Monvernay 1999) and in determination of *trans*-fatty acids in hydrogenated oils (Di Oliviera et al. 2003). Further, Lerma-García (2007) have developed a method based on capillary electrophoresis for the analysis of vegetable oils for tocopherols, using methacrylate ester-based monolithic columns.

3.10 Conclusion

The present review shows different analytical methods used for the analysis of lipid components. The choice of the best method depends, mainly, on the scope of the analytical control, the amount of information that can be acquired and the cost of the overall analytical operation.

References

Al-Ismail, K., Alsaed, A.K., Ahmad, R. and Al-Dabbas, M. 2010. Detection of olive oil adulteration with some plant oils by GLC analysis of sterols using polar column. Food Chem. 121: 1255–1259.
Avalli, A. and Contarini, G. 2005. Determination of phospholipids in dairy products by SPE/HPLC/ELSD. J. Chromatogr. A. 1071: 185–190.
Azadmard-Damirchi, S. and Dutta, P.C. 2006. Novel solid-phase extraction method to separate 4-desmethyl-, 4-monomethyl-, and 4,4-dimethylsterols in vegetable oils. J. Chromatogr. A. 1108: 183–187.
Baeten, V., Hourant, P., Morales, M.T. and Aparicio, R. 1998. Oil and fat classification by FT-Raman spectroscopy. 46: 2638–2646.
Baeten, V. and Aparicio, R. 2000. Edible oils and fats authentication by Fourier transform Raman spectrometry. Biotechnologie, Agronomie, Société et Environnement (BASE). 4: 196–203.
Baeten, V., Dardenne, P., Meurens, M. and Aparicio, R. 2001. Interpretation of Fourier transform Raman spectra of the unsaponifiable matter in a selection of edible oils. J. Agric. Food Chem. 49: 5098–5107.

Baeten, V., Fernandez Pierna, J.A., Dardenne, P., Meurens, M., Garcia Gonzalez, D.L. and Aparicio, R. 2005. FT-IR and FT-Raman spectroscopy and Chemometric techniques for the analysis of olive and hazelnut oils. J. Agric. Food Chem. 53: 6201–6206.
Baeten, V. 2010. Raman spectroscopy in lipid analysis. Lipid Technol. 22: 36–38.
Beattie, J.R., Bell, S.E.J., Borgaard, C., Fearon, A.M. and Moss, B.W. 2004. Multivariate prediction of clarified butter composition using Raman spectroscopy. Lipids. 39: 897–906.
Benitez-Sanchez, P.L., Leon-Camacho, M. and Aparicio, R. 2003. A comprehensive study of hazelnut oil composition with comparisons to other vegetable oils, particularly olive oil. Eur. Food Res. Technol. 218: 13–19.
Bligh, E.G. and Dyer, W.J. 1959. A rapid method of total lipid extraction and purification. Can. J. Bioch. Physiol. 37: 911–917.
Cert, A., Moreda, W. and Perez-Camino, M.C. 2000. Chromatographic analysis of minor constituents in vegetable oils. J. Chromatogr. A. 881: 131–148.
Christie, W.W. 1989. Gas chromatography and Lipids. Oily press Ltd. Scotland.
De Oliveira, M.A.L., Solis, V.E.A., Gioielli, L.A., Polakiewicz, B. and Tavares, M.F.M. 2003. Method development for the analysis of *trans*-fatty acids in hydrogenated oils by capillary electrophoresis. Electrophoresis. 24: 1641–1647.
Destaillats, F., Cruz-Hernandez, C., Nagy, K. and Dionisi, F. 2010. Identification of monoacylglycerolregio-isomers by gas chromatography–mass spectrometry. J. Chromatogr. A. 1217: 1543–1548.
Dillon, J.T., Aponte, J.C., Tarozo, R. and Huang, Y. 2012. Efficient HPLC analysis of Mono-, Di-, and Triglycerols using silver Thiolate chromatographic material. J. Chromatogr A. 1240: 90–95.
Fanny, A., Abert-Via, M., Peltier, G. and Chemat, F. 2012. "Solvent-free" ultrasound-assisted extraction of lipids from fresh microalgae cells: A green, clean and scalable process. Bioresource Technology. 114: 457–465.
Folch, J., Lees, M. and Sloane Stanley, G.M. 1957. A simple method for the isolation and purification of total lipids from animal tissues. J. Biol. Chem. 226: 497–509.
Fuchs, B., SuB, R., Teuber, K. and Eibisch, M. 2011. Lipid analysis by thin-layer chromatography—A review of the current state. J. Chromatogr A. 1218: 2754–2774.
Fuchs, B., Sus, R. and Schiller, J. 2010. An update of MALDI-TOF mass spectrometry in lipid research. Prog. Lipid Res. 49: 450–475.
Guichard, C. 1967. Elements de Pharmacie et de Technologie Pharmaceutique (Pharmacie Galenique), Flammarion.
Guo, Q., Xu, G. and Chang, L. 1998. Analysis of soybean phospholipids by Thin layer chromatography. Chinese J. Anal. Chem. 26: 81.
Herchi, W., Harrabi, S., Rochut, S., Boukhchina, S., Kallel, H. and Pepe, C. 2009a. Characterization and Quantification of the Aliphatic Hydrocarbon Fraction during Linseed Development (*Linum usitatissimum* L.). J. Agr. Food Chem. 57: 5832–5836.
Herchi, W., Harrabi, S., Sebei, K., Rochut, S., Boukhchina, S., Pepe, C. and Kallel, H. 2009b. Phytosterols accumulation in the seeds of *Linum usitatissimum* L. Plant Physiol. Biochem. 47: 880–885.
Hidalgo, F.J. and Zamora, R. 2003. Edible oil analysis by high-resolution nuclear magnetic resonance spectroscopy: recent advances and future perspectives. Trends Food Sci. Technol. 14: 499–506.
Hopkins, C.Y. 1961. Nuclear magnetic resonance in lipid analysis. J. Am. Oil Chem. Soc. 38: 664–668.
Horn, P.J., Ledbetter, R.N., Christopher, N.J., Hoffman, W.D., Case, C.R., Verbeck, G.F. and Chapman, K.D. 2011. Visualization of lipid droplet composition by direct organelle Mass Spectrometry. J. Biol. Chem. 286: 3298–3306.
Jin, F., Kawasaki, K., Kishida, H., Tohji, K., Moriya, T. and Enomoto, H. 2007. NMR spectroscopic study on methanolysis reaction of vegetable oil. Fuel. 86: 1201–1207.

Knothe, G. and Kenar, J.A. 2004. Determination of the fatty acid profile by ^{1}H-NMR spectroscopy. Eur. J. Lipid Sc. Technol. 106: 88–96.
Korostynska, O., Blakey, R., Mason, A. and Al Shamma'a, A. 2013. Novel method for vegetable oil type verification based on real-time microwave sensing. Sensors and Actuators A: Physical. 202: 211–216.
Lange, R. and Fiebig, H.J. 1999. Fett /Lipid. 101: 77.
Lerma-García, M.J., Simó-Alfonso, E.F., Ramis-Ramos, G. and Herrero-Martínez, J.M. 2007. Determination of tocopherols in vegetable oils by CEC using methacrylate ester-based monolithic columns. Electrophoresis. 22: 4128–4135.
Matsufuji, H., Nakamura, H., Chino, M. and Takedaj, M. 1998. Antioxidant activity of capsanthin and the fatty acid esters in paprika (*capsicum annum*). Agric.Food Chem. 46: 3468.
Mofaddel, N. and Desbène-Monvernay, A. 1999. L'analyse des acides gras en électrophorèse capillaire. Analusis. 27: 120–124.
Muik, B., Lendl, B., Molina-Daz, A. and Ayora-Caada, M.J. 2005. Direct monitoring of lipid oxidation in edible oils by Fourier transform Raman spectroscopy. Chem. Phys. Lipids. 134: 173–182.
Naudet, M. 1992. Principaux Constituants des Corps Gras, in Manuel des Corps Gras, Tech. & Doc. Lavoisier, tome (I), pp. 65–94.
Nguyen, H. and Tarndjiiska, R. 1995. Lipid classes, fatty acids and triglycrides in papaya seed oil. Lipid/Fett. 97: 20–23.
Pulfer, M. and Murphy, R.C. 2003. Electrospray mass spectrometry of phospholipids. Mass Spec. Rev. 22: 332–364.
Rao, K.S., Chakrabarti, P.P., Rao, B.V.S.K. and Prasad, R.B.N. 2009. Phospholipid composition of Jatrophacurcus seed lipids. J. Am. Oil Chem. Soc. 86: 197–200.
Santos, J.C.O., Santos, M.G.O., Dantas, J.P., Conceicao, M.M., Athaide-Filho, P.F. and Souza, A.G. 2005. Comparative study of specific heat capacities of some vegetable oils obtained by DSC and microwave oven. J. Therm. Anal. Calorim. 79: 283–287.
Saraiva, S.A., Cabral, E.C., Eberlin, M.N. and Catharino, R.R. 2009. Amazonian vegetable oils and fats: Fast typification and quality control via Triacylglycerol (TAG) profiles from dry Matrix-Assisted Laser Desorption/Ionization Time-of-Flight (MALDI-TOF) Mass Spectrometry fingerprinting. J. Agric. Food Chem. 57: 4030–4034.
Tlili, N., Trabelsi, H., Renaud, J., Khaldi, A., Mayer, P.M. and Triki, S. 2011. Triacylglycerols and phospholipids composition of Caper Seeds (*Capparis spinosa*). J. Am. Oil Chem. Soc. 88: 1787–1793.
Trabelsi, H., Cherif, O.A., Sakouhi, F., Villeneuve, P., Renaud, J., Barouh, N., Boukhchina, S. and Mayer, P. 2012. Total lipid content, fatty acids and 4-desmethylsterols accumulation in developing fruit of *Pistacia lentiscus* L. growing wild in Tunisia. Food Chem. 131: 434–440.
Trabelsi, H., Renaud, J., Herchi, W., Khouja, M.L., Boukchina, S. and Mayer, P. 2013. LC-ESI-QTOF-MS, MS/MS analysis of glycerophospholipid species in three Tunisian *Pistacia lentiscus* fruit populations. J. Am. Oil Chem. Soc. 90: 611–618.
Tranchida, P.Q., Salivo, S., Franchina, F.A., Bonaccorsi, I., Dugo, P. and Mondello, L. 2013. Qualitative and quantitative analysis of the unsaponifiable fraction of vegetable oils by using comprehensive 2D GC with dual MS/FID detection. Anal. Bioanal. Chem. 405: 4655–4663.
Wewer, V., Dombrink, I., vom Dorp, K. and Dörmann, P. 2011. Quantification of sterol lipids in plants by quadrupole time-of-flight mass spectrometry. J. Lipid Res. 52: 1039–1054.
Wong, M.L., Timms, R.E. and Goh, E.M. 1988. Colorimetric determination of total tocopherols in palm oil, olein and stearin. J. Am. Oil. Chem. Soc. 65: 258–261.

Zábranská, M., Vrkoslav, V., Sobotníková, J. and Cvačka, J. 2012. Analysis of plant galactolipids by reversed-phase high-performance liquid chromatography/mass spectrometry with accurate mass measurement. Chemistry and Physics of Lipids. 165: 601–607.

Zaima, N., Goto-Inoue, N., Adachi, K. and Setou, M. 2011. Selective analysis of lipids by thin-layer chromatography blot matrix-assisted laser desorption/ionization imaging mass spectrometry. J. Oleo Sc. 60: 93–98.

4

Texture in Meat and Fish Products

Purificación García-Segovia,* Mª Jesús Pagán Moreno and Javier Martínez-Monzó

ABSTRACT

Meat and fish are complex biological materials, which are heterogeneous and anisotropic. It is therefore difficult to carry out fundamental rheological measurements on such products. In addition, food scientists are often interested in complex sensory properties, such as the tenderness or chewiness of a meat or fish product, with contributions of both shear and compression, and usually involving large deformations. For this reason, tests on meat or fish are often carried out using empirical instruments. For example, a device has been developed that measures the force required for a blade to slice through a piece of meat or a thin blade shear/compression cell for fish.

However, many common derivate products from meat and fish (such as restructured meat or sausages) exhibit complex rheological properties, with viscosity and viscoelasticity that can vary depending upon the external conditions applied, such as stress, strain, timescale and temperature. Internal sample variations can also be key factors, such as protein concentration and hydrocolloids present in their formulation. Rheological properties are critical as they impact at all stages in meat and fish industries—from formulation development and stability to processing to product performance.

Department of Food Technology, Universitat Politécnica de Valencia, Valencia, Spain.
* Corresponding author

This chapter reviews different methods to measure texture and rheological properties in meat and fish products.

4.1 Introduction

Rheology and texture are important properties that are affected during the processing of food in production plants. During this processing, significant changes occur in the composition and structure of meat and fish products, influencing quality, appearance and appeal of products and impacting the mouth feel and organoleptic qualities (García-Segovia et al. 2011).

Knowledge of textural properties of ingredients used in meat and fish products is crucial for successful new developments. Viscosity, thickness, hardness, firmness, chewiness, toughness, stringiness, and brittleness (fracturability) are some of the basic characteristics that are associated with the physical properties of meat and fish products that can have the greatest influence on consumer acceptance (García-Segovia et al. 2011).

Rheology is the study of flow and deformation of materials under applied forces. The measurement of rheological properties is applicable to all material types—from fluids such as dilute solutions of polymers and surfactants through to concentrated protein formulations, to semi-solids such as meat or fish pastes (Steffe 1996).

A general agreement has been reached on the definition of texture as *"the sensory and functional manifestation of the structural, mechanical and surface properties of foods detected through the senses of vision, hearing, touch and kinesthetic"* (Szczesniak 2002).

The purpose of this chapter is to review the practical methods to measure texture properties in meat and fish and their derivate products to understand the changes in the structure of these products.

4.2 Measuring Texture: The Basis of Test Methods

Some of the tests most commonly used for characterizing textural properties of meat and fish and their derived products were described by García-Segovia et al. (2011). Instrumental texture assessment can be made by means of different commercial texturometers (Instron, TA.XT2i Stable Microsystems Texture Analyzer, TMS from Food Technology Corporation, or TA1 from Ametek). These systems provide a number of cells that allow tissues to be evaluated using different methods. Table 4.1 categorizes different cells developed, test methods used and attributes measured in meat or fish products.

Table 4.1. Cells in texture analyzers.

Cells	Test Methods
Knife blade	Shearing test
Standard Warner-Bratzler	Shearing test
European Warner-Bratzler	Shearing test
Volodkevich Bite Jaws	Smaller shearing test
Razor blade	Shearing test
5-blade Kramer Cell	Bulk analysis: Multiple shearing test
Ottawa Cell	Bulk analysis: Extrusion test
Needle Probes	Penetration test
Multiple Puncture Probe	Multiple point test
Compression Platens	TPA test
Back extrusion fixture	Back extrusion
1" Ball Probe	"Finger" method
Tensile Grips	"Hold Until Time" test

4.3 Guidelines for Measuring Meat and Fish Texture

Usually foods compete for consumer preference by innovation of new products and flavors, whereas in meat, fish and poultry products it is the texture which is the main factor that is perceived as quality by consumers. Should the texture, specifically tenderness, be too tough or too tender, or differ from expectations in any other way, the product perception may be strongly affected.

A number of texture test methods have been developed in the last decades for measuring texture of food products. The best advice is to make texture testing practical and real, by using objective techniques that best replicate handling by the consumer with sensory analysis.

As whole-tissue meat and fish are natural products, possessing inherent variability due to many factors, the common problem is to assess the texture reproducibly. In the case of processed meats, because they are of different types, the measurement issues also vary. Based on the texture measurements, Food Technology Corporation (FTC 2012) classifies meat products into:

a) **Pastes and viscous liquids** (meat slurries, pastes, potted meats, terrines, etc.): semi-solid products with weak gelly structure. These products are supplied in containers due to unsupported structure. They may contain particulates such as onions, coarse muscle fibres and nuts, and the food system may be emulsified to support particulates.

b) **Particulates** (minced and diced meats, reconstituted meat pieces, etc.): small, irregular and non-uniform particulate pieces with fibrous solid structure. The tests measurements will quantify the toughest

components (including those that carry gel inside) providing an indication of firmness from shearing or puncture.
c) **Homogeneous solids** (pâté, mousses, liver sausage, meat jellies, hotdogs and frankfurters, etc.): smooth viscous pastes or homogeneous gelled products with uniform yield stress. Highly elastic when lightly squeezed, but fail once maximum resistance is reached. Generally they fracture, split or begin to flow, once yield force or conditions are reached.
d) **Inhomogeneous solids** (hamburgers, sausages, reconstituted cooked ham, salami, etc.): inhomogeneous solid products made of suspended particulates of varying size and shape. Often incorporate gelling or binding components to crate self-supporting and emulsified structure. Presence of gelling agents creates elastic structure at small deformations. Products will fail or collapse when squashed to higher levels, reverting to particulate or component structure.
e) **Fibrous solids (whole muscles)**: bundles of muscle fibres tightly grouped together. Multiple groups of fibre bundles are often found within a single sample, which can result in combinations of fibre orientations. Result reproducibility is improved when single orientation of fibres is used in a traditional shear test. If muscle tissue is in variable form, multiple-point analysis using Kramer shear cell is advised to measure the toughest component.

The characteristics of these products are different from semi-solid (a) to solid (e) structure and their treatments during handling and consumption also differ. The texture parameters and the type of analyses to realize are conditioned by all of these factors. The semi-solid products are poured, pumped, extruded or spread and the solid products deformed, squashed, sheared or snapped during handling or consumption. For homogeneous products, solid firmness and extensibility, performance and flow point, gel strength and point of rest, relaxation and gel strength, elasticity and resiliency are important parameters. In case of inhomogeneous solids, slicing resilience, hardness to bite (shearing), hardness to touch (compression), elasticity and consistency comparison, bond strength of emulsion, sensory comparisons and objective measures as skin peel strength are important. Finally, in fibrous solids the most important determinations are muscle fibre toughness, bonding strength between fibres, shear toughness and firmness of muscle, elasticity and resistance to chewing, maximum bite resistance and tensile strength and bonding (FTC 2012).

In order to assess texture parameters in meat, fish and poultry products, the most common methods employed are: shearing, penetration and compression. Other methods including tension, back extrusion, bulk analysis, and multiple point analysis are also used.

In the following paragraphs these methods are described and some practical applications shown.

4.3.1 Shearing Test

Food shearing testing is frequently used in food texture analysis. It is very common since many foods are first sliced or "sheared" by the incisors when introduced into the mouth. Consequently, product texture variations can be measured by slicing through the whole sample with blades imitating the action of the incisor teeth to assess meat properties. Depending on the blade geometry, different actions are performed on the sample, including shearing, tearing and compression. When low force measurements are made, a simple wire blade may be used, for example for a raw meatball, whereas a shear blade is required for heavier-duty tests, such as shear testing of cooked chicken breast to measure toughness. Attributes assessed by the shearing test include bite strength, firmness, tenderness and toughness.

There are many different variations on the basic shear testing apparatus that may use razor blades, V-shaped blades, rounded blades and straight blades. The validity and repeatability of the measurements of shear depend on both instrumentation and methodology (Melito and Daubert 2011).

Warner-Bratzler

The most common shearing system for the assessment of meat, poultry and fish products is the Warner-Bratzler (WB) shear blade (Bratzler 1932; 1949; Warner 1928). This empirical technique developed 70 years ago remains the main objective reference for the assessment of raw and cooked products (Rees et al. 2002; Oeckel et al. 1999; Xiong et al. 2006). The most commonly used configuration is the one in which the shearing plane is perpendicular to the muscle fibres. Tensile, shear and compression forces operate in this test. The WB technique usually yields the best correlation with the sensory panel scores for meat toughness (Cover et al. 1962).

Figure 4.1 shows the typical deformation curve (force exerted versus time or distance) obtained in a WB test. The sample should be sheared at right angles to the fibre axis. The parameters to be measured are the maximum shear force (the highest point of the curve) (N or kg) that indicates the maximum resistance of the sample to shearing and the total energy (N x mm) (the total work necessary for the total sample cut) (Honnikel 1998).

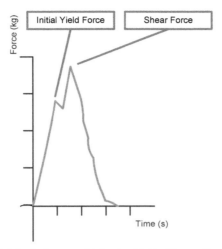

Figure 4.1 A typical curve obtained with the Warner-Bratzler cell.

Voisey (1976) indicated that the simplicity of the WB device is associated with a complex interpretation of the data curve. Several studies examined the appearance of the shear deformation curve and show that the WB test can provide information about the two structural components of meat, myofibrillar protein and connective tissue. Usually, the small peaks before the shear force maximum are associated with the myofibrillar components and correspond to the initial yield while subsequent peaks correspond to the connective tissue (Bourton and Harris 1978; Honikel 1998; Möller 1980; Möller et al. 1981; Girard et al. 2011; Girard et al. 2012a; Girard et al. 2012b). Barroso et al. (1998) using an interpretation similar to the one used for meat (Möller 1980; 1981), noted for the fish muscle that the first peak of the force-deformation curve is due to the muscle, while the second, much sharper and larger peak, is due to the connective tissue. In fish, the Warner-Bratzler test gives good information of the shear-strength, but it is necessary to act very carefully and precisely as the measurement is done on a small shear area and hence there can be too much variability between replicates, depending on the number and direction of the myotomes and myosepts being cut, especially in species of small size (Montero and Borderias 1990; 1992; Chamberlain et al. 1993).

Factors that affect the results of the Warner Bratzler shear test are: uniformity of sample size, direction of muscle fibres, presence of connective tissue and fat deposits, sample temperature, and speed of shearing. Normally the WB measurement is performed according to the method proposed by Honikel (1998) with modifications. The WB test results depend of the sample, storage of samples, cooking and testing (equipment, procedure and evaluation). Therefore, the sampling location and characteristics must be clearly described. It is recommended to use a slice of muscle perpendicular to the longitudinal axis of the muscle and thick enough to produce muscle fiber lengths of least 50 mm along the fiber axis. Ideally, assessments should be performed immediately after sampling. For frozen product the conditions of freezing must be specified because these will affect the tenderness measurement and ideally cooking should be carried out from the frozen state. If thawing is necessary, this must be specified, as the freeze/thaw cycles may affect the tenderness of the samples. The temperature-time conditions of cooking are important on sample force determination. Connective tissue effects predominate at cooking temperatures up to 60 °C, while above that myofibrillar components become more important. For this reason, the end-point temperature in the center of the sample must be measured, 75 °C is recommended. After cooking the samples should be cooled and may be stored for up to four days in chilling conditions (1–5 °C).

The basic concept and design of the WB shear device has been subject to modification and improvement over the years. Originally it was used on meat products. However, blade set applications now include sausage and other products that require testing by cutting or shearing. Therefore, WB devices and sample configuration are extremely variable. The WB shear methods have been implemented in multiple instruments for meat texture evaluation including the Texture Measurement Systems (Food Technology Corp., Slinfold, U.K.), the Texture Technologies Texture Analyzer (TA) (Texture Technologies Corp., Scarsdale, NY) and the Instron Universal Testing Machine (Instron Corp., Canton, MA). Figure 4.2 shows different Warner Bratzler tests.

The standard WB test cell has a 3 mm-thick steel blade with a 73° V notch in its lower edge (Fig. 4.2a), going through a 4 mm wide slit in a small table (like a guillotine with a V cut into the blade). The USDA standard WB shear force test involves measurement of cooked meat tenderness using a WB blade with the following specifications: a) shearing blade thickness of 1.016 mm; b) a V-shaped notch with a 60° angle; c) the cutting edge beveled to a semi circle; d) the corner of the V should be rounded to an arc of a 2.363 mm diameter circle; e) the spacers providing the gap for the cutting blade to slide through should be 1.245 mm thick; f) the cooked meat samples should be round cores 1.27 cm in diameter removed parallel to the longitudinal orientation of the muscle fibres; and g) the cores should be sheared once at the centre, perpendicular to the fibres to avoid hardening that occurs toward the surface of the sample (Bratzler 1932; 1949; AMSA 1995; Wheeler et al. 1995). The maximum force is termed the WB shear force. This configuration can be used on raw or cooked muscle meat, including beef, lamb, pork, poultry, processed meat, sausages, and fish, etc. Sometimes, it is considered

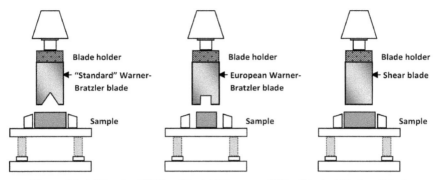

Figure 4.2 a) "Standard" WB cell test, b) European WB cell test, c) Blade cell test.

that tests conducted with modifications to these specifications (e.g., square notch in the blade, square meat samples, straight cutting blade, or blade edge not beveled) should not be referred to as the WB shear force tests.

The width of the blade and the location of the triangle, the speed of the blade, the shape, mass, and orientation of the test sample are of course important for interpretation of the test results. The standard methods and attachments have been modified by groups of researchers who prefer to move away from a standard to a method that suits their particular purposes. One such group is responsible for the development of a new variant of the original WB blade. As a spin-off of a workshop on pork quality, held in Helsinki in 1992, a group of scientists with many years of experience in the field of meat quality assessment convened in 1993 for the first time, and subsequently in 1994 and 1995 in Kulmbach at the German Federal Centre for Meat Research, to develop internationally accepted reference methods. In the autumn of 1997 these methods (including a method for tenderness) were brought into their final form at the Meat Industry Research Institute of New Zealand. The recommended equipment according to this "European" network is as follows (Fig. 4.2b). The blade should be 1.2 mm thick with a rectangular notch (11 mm wide and at least 15 mm high). The notch should have square smooth edges and the blade should be drawn or pushed at 50.0 mm/min between the side plates positioned to provide a minimum gap to the blade. The sample should be cut from a block of cooked meat in a way to avoid damage. Sample strips should be cut with a 100 mm^2 (10 mm x 10 mm) cross-section with the fibre direction parallel to a long dimension of at least 30 mm.

Finally, the test cell (Fig. 4.2c) may come with a reversible blade, a slotted blade and a blade holder. The reversible blade has a knife edge at one end and a flat guillotine edge at the other. In operation, the blade is firmly held in the blade holder. The slotted blade insert is installed directly into the heavy-duty platform, acting as a guide for the blade whilst providing support for the product.

The WB cell is one of the most common devices used in the texture analysis in meat, poultry and fish. Girard et al. (2012b) studied the contribution of myofibrillar and connective tissue components to the WB shear force of cooked beef. In this case, the muscles analyzed were the semitendinosus and gluteus medius. Thick steaks of 2.5 cm were cut from each muscle and grilled at 210 °C until the internal temperature of 71 °C. Then the steaks were cooled at 4 °C. The samples used in the WB test were cores (1.9 cm in diameter) removed from cooked steaks in parallel to the muscle fibre direction. The shear force was measured perpendicular to the muscle fibres using a Warner Shear head (Texture Technologies Corp, New York) at a crosshead speed of 200 mm/min. The results obtained showed correlations between shear forces and meat quality. Ahnstrom et al. (2006; 2012) used the European WB cell test in studies of the effect upon textural properties of pelvic and conventional hanging by the Achilles tendon in bull, heifer, and cow carcasses. Five muscles (longissimus dorsi, semimembranosus, adductor, psoas major, and gluteus medius) were analyzed. Tenderness was measured by the WB shear force method described by Honikel (1998) with modifications. From each sample, 40–30 mm × 10 mm × 10 mm strips (a shear area of 100 mm^2) were removed with the fibre direction parallel to the longitudinal dimension of the strip. The samples were cooked and stored at 4 °C before shearing at room temperature. A minimum of 8 and maximum of 16 strips were sheared perpendicular to the fibre axis. The blade had a cross-head speed of 0.83 mm/s when cutting. The results obtained showed that the effect of pelvic suspension merits industrial consideration because it improves muscle yields, tenderness, and reduces variation within muscles. Lyon and Lyon (1998) in their study "Assessment of three devices used in shear tests of cooked breast meat" used three shearing methods: the Benchtop WB (BT-WB) machine, the WB blade attachment (TA-WB) and a 45° chisel-edge blade attachment (TA-WD). The TA-WB and TA-WD were attached to Model TA.XT2 texture analyzer. The BT-WB standard shear device (Model 3000) was fitted with a 1.15-mm thick stainless-steel blade with a triangular notch, the speed was 4 to 5 mm/s. The capacity spring force was 25 kg. In the TA-WB the blade consisted of a rectangular notch with an inverted V and measured 70 mm wide, 90 mm high, and 3.2 mm thick. The cell load capacity was 50 kg and the blade traveled at 5 mm/s. In the TA-WD, the device was a TA-XT2 special attachment (TA-42) with a 45° chiseled edge. The dimensions of the blade and the cross arm travel were the same as for BT-WB. After different treatments (electrical stimulation, post-mortem deboning times and marination) and before the textural tests, the samples were cooked (to 80 °C in the core). This study indicated that the TA-WB was the least sensitive in detecting differences in shear values due to treatment. The shear values obtained were from 3.68 kg to 15.37 kg, depending on the treatment and the device used. Zhuang and Savage (2009) evaluated the effects of deboning aging on poultry tenderness using WB test. This was performed on samples cooked in vacuum bags at 85 °C to an internal temperature of 78 °C. Three 1.9 cm wide strips were removed from the breast and each sample sheared using a TA-XTPlus Texture Analyzer and a TA-7 WB shear-type blade. The results demonstrated that the variation of WB shear force measurements and the Pearson correlation coefficients may be affected by postmortem aging time and storage.

Combes et al. (2010) analyzed the influence of cage or pen housing on carcass traits and meat quality of rabbit. They used the European WB test to evaluate the texture of rabbit meat and the samples were prepared according to Honikel (1998). The WB shear test was made using a universal test machine (Synergie 200; MTS, Eden Prairie, MN, USA). The samples were cut with a 100 mm^2 cross-section with the fibre direction parallel to the long dimension of 2 cm from abductor cruralis cranialis, biceps femoris and semitendinosus muscles. The samples were sheared at right angles to fibre axis using a WB shear blade with a rectangular notch and a speed of 100 mm/min. In this case, the parameters measured were the maximum shear force (N), the energy (mJ) and the stiffness (ratio of maximum shear force to displacement at maximum shear force, N/mm). The experiment indicated that large pen housing induce modifications in carcass and meat quality that are in line with consumers' demand. In ham, WB test was used to evaluate the rheological characteristics of sliced vacuum-packaged dry-cured ham, submitted to high-pressure treatments (Alba et al. 2012), using a 4301 Compression Tester (Instron Ltd., Barcelona, Spain) with a load cell of 1000 N and crosshead speed of 100 mm/min. For the texture analyses the dry cured ham was sliced into 4-mm-thick slices. Guerrero et al. (1999) studied the influence of meat pH (normal and DFD) on mechanical and sensory textural properties of dry-cured ham. In this study different tests were used to evaluate the textural properties of ham (WB shear test, uniaxial compression test, puncture test and texture profile analysis or TPA). The WB test was performed using a triangular-notch cutting edge (Instron Universal Testing Machine model 4301). The test was done on samples of 1x1 cm and 2 cm along the fibre axis with crosshead speed of 50 mm/min and the distance travelled by the shear blade equal to the thickness of the ham sample. From the WB force-distance curve three parameters were measured: initial yield force (kg), maximum force (kg) (Moller 1980) and shear firmness (kg/min) (Brady and Hunecke 1985). The results indicated greater hardness in the normal hams, in accordance with the sensorial results. However, only the yield force measured with the WB device was significantly different in the two types of hams. This parameter is in fact related to the myofibrillar components of the muscles (Moller 1980). The results indicate that the final meat pH affects the components of the muscle. The WB has also been used to evaluate the tenderness of different types of sausage. Caceres et al. (2004; 2006) studied the effect of enrichment with calcium and fructooligosaccharides on the sensory characteristics of cooked sausage. In this study, the texture properties were evaluated by WB shear test, texture profile analysis and Kramer shear using TA.XT 2i/25 texturometers (Stable Micro System, London, UK) at room temperature. In the WB shear test the samples were cores of 1 cm thick and 2.5 cm diameter and the crosshead speed was 2 mm/s. The results recorded were shear force (N) and work done to move the blade (N.s). The results showed values of 5.7–9.2 (N) for shear force and 37.5–62.6 (N.s) for shear work. The authors reported a slight tendency toward an increase in the shear work with the addition of salts (Caceres et al. 2006) and a decrease in this parameter with the addition of soluble dietetic fibre (Caceres et al. 2004).

A shearing method similar to the WB has been reported by Chamberlain et al. (1993). A fish-shearing device (FSD) uses a blade, which cuts the sample as it traverses a rectangular or circular device. Other devices similar to WB have been developed (Kanoh et al. 1988). When the WB test is used in fish it is more important to take account of the orientation of the fibres with the blade and the relationship with diameter or cross-section of the test samples because it is difficult to test small pieces of fish muscle (Barroso et al. 1998). Another problem is that the cell requires frequent dismounting for cleaning and recalibration (Chung and Merrit 1991). Nevertheless, different studies report that the shear test is slightly more sensitive in some applications than the texture profile analysis (TPA) (Veland and Torrissen 1999; Sigurgisladottir et al. 1999). Christensen et al. (2011) used the WB test and TPA to evaluate textural changes during ripening of traditionally salted herring. The maximum shear force was measured on an Instron texturometer equipped with a load cell of 100 N and a triangular WB shear blade. The cross-head speed was 50 mm min^{-1}. Initially, the herrings were filleted and cut parallel to fibre direction (fillets with approximately 10 mm in width) and stored on ice. Before the analysis, the exact height and width of samples were determined. The shear force values were from 65 to 125 kPa depending of the characteristics of the samples. Veland and Torrissen (1999) used two different tests to determine the texture of Atlantic salmon (*Salmo salar*). Both tests performed well, but the shear test was slightly more sensitive than the compression test. The measurements were performed using the TA-XT2 texture analyzer (Stable Micro Systems, Surrey, England) in WB test; samples (25 mm diameter) were cut perpendicular to the plane of the fillet. The standard WB cell was used. The cross head speeds were 5 or 2 mm s^{-1} (depending on the experiment) and shearing force measured every 0.01 s. The WB test was also used on cod (Fernandez-Segovia 2003) using 20 mm penetration at a constant deformation rate of 1 mm/s. The sheer force values were 24.5 N in desalted cod without thermal treatment and 2–3.6 N in samples with thermal treatment. Other studies used the WB test, for example in restructured adductor muscle of the Pacific calico scallop (Beltran-Lugo et al. 2005), Australian red claw crayfish (Tseng et al. 2002), fish fingers (Schurbring 2000), snack foods of pink salmon (Choudhury et al. 1998; Gautam et al. 1997) and dried atka mack (Iseya et al. 1996). It is interesting to use the WB test to evaluate texture properties in restructured fish products (Uresti et al. 2003; Perez-Won et al. 2006; Cardoso et al. 2010). Several authors indicate a good correlation between the results of shear test and toughness (Chamberlain et al. 1993), WB values and hardness (Morkore and Einen 2003) or firmness (Pérez-Won et al. 2006).

Razor Blade

The craft knife or razor blade test is carried out using a replaceable blade which is attached to the instrument (Fig. 4.3a). With this test hard samples can be sheared with great precision without compressing (Henrickson et al. 1974). Hardness is determined by the maximum force recorded

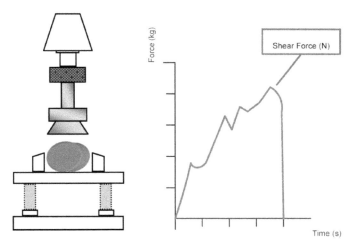

Figure 4.3 a) Razor blade or craft knife and b) Typical curve.

during the test (N) and the energy (N.mm) is calculated as the area under the force-deformation curve (from the beginning to the end of the test) (Fig. 4.3b). The brittleness can also be determined from the fracturability values. Almond crunchiness is determined from the measurement of the quantity of fractures generated during the test.

Recently, an instrumental method has been developed by The University of Arkansas for quantifying poultry firmness (Fig. 4.4), the Meullenet-Owens razor shear (MO, RB, RBS, MOR, MORBS or MORS) having been established as a reliable predictor (Cavitt et al. 2001). Several studies reported that MORS has a high correlation to consumer sensory attributes (hardness and cohesiveness) (Cavitt et al. 2001; 2004; 2005a,b; Meullenet et al. 2004; Xiong et al. 2006). Also, it has some advantages over other devices as the WB or Kramer cell, being less time-consuming, simpler to perform and equivalent in performance to the WB and Kramer shear in evaluating poultry breast meat tenderness. In the MORS, a razor blade (8.9 mm width and 20 mm high) penetrates 20 mm into the samples using 5 kg load cell. The crosshead speed should be 10 mm/s and triggered by a 10 g contact force. The maximum shear force (N) and energy (N.mm) values are obtained. Lee et al. (2008) in their study of applications and optimization of the MORS for evaluating poultry meat tenderness indicated that the use of intact samples minimizes experimental errors.

A blunt version of the MORS (BMORS) was developed with the intention of providing better discrimination among tough cuts of meat. In addition, BMORS solves the problem of having to change shearing blades every 100 shears (Meullenet et al. 2004). Lee et al. (2008) evaluated the effectiveness and reliability of the MORS and BMORS for evaluating tenderness of broiler breast meat. Tenderness of breast fillets which had been frozen, thawed

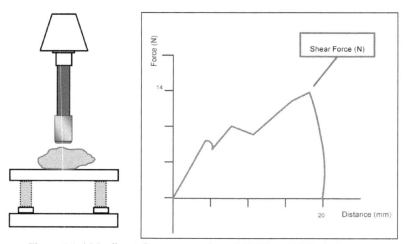

Figure 4.4 a) Meullenet-Owens razor shear (MORS) and b) Typical curve.

overnight, and then cooked in a convection oven to an internal temperature of 76 °C was evaluated using several different texture analyzers. Both methods were equivalent in performance, although the values of BMORS energy were better correlated to the tenderness perceived by consumers. The MORS has been utilized in poultry meat and the applications of the MORS or BMORS to other meats are few and require further study.

Zhang et al. (2013) measured the shear force in duck using a texture analyzer (TVT-300XP, TexVol Instruments, Viken, Sweden) equipped with a razor blade (24 mm x 8.9 mm). The crosshead speed was at 2 mm/s and the test was triggered by a 10 g contact force according to Cavitt et al. (2005b). This shear blade test was utilized for studying the effect of frying in poultry lean meat portions. The blade was mounted in a texture analyzer (Model TA.XT2, Stable Micro Systems, Texture Technologies Corp., Scarsdale, NY) and moved down 15 mm, at 1.5 mm/s (Barbut 2013). Harper et al. (2012) adapted the razor blade method (Cavitt et al. 2005a,b; Houben et al. 2005) to the textural study of manufactured collagen sausage casings. Uncooked and cooked sausages were sheared (lengthwise and widthwise) using a 9 mm craft knife and using a texture analyzer (Model TA.XT2, Texture Technologies Corp., Scarsdale, NY, USA) with a 30 kg load cell at a crosshead speed of 1.5 mm/s. The trigger force was set at 0.049 N and the penetration depth was 15 mm. The shear force (N) and shear distance at break (mm) of the casings were determined using the first peak in the force–distance graph. The work (N x mm) to shear was calculated as the resultant area under the force-distance curve to the first peak. The initial slope of the force-distance curve (N/mm) was also determined. Sausages were tested at room temperature (21 °C). Sawyer et al. (2007) used the MORS to evaluate the tenderness in beef. In this case, steaks (1.91 cm thick) were cooked until internal temperature of 71 °C and left to cool at 22 °C. The test was performed with a 49 N load cell, a crosshead speed of 10 mm/s, 0.1 N of contact force and blade penetration depth of 20 mm. The device used was Texture Analyzer (Model TA-XT2i; Texture Technologies, Scarsdale, NY, USA). Baublits et al. (2006) found that the MORS was a reliable measure of cooked pork tenderness. Samples of pork chops (2.54 cm thick) were cooked. After, steaks were allowed to cool to room temperature (22 °C) for approximately 2 h. The force required to shear perpendicular to the transversely cut surface of each steak was determined using a 49-N load cell and a blunt-blade shear attachment (height = 24 mm and width = 8 mm). The crosshead speed was 10 mm/s, 0.1 N of contact force initiated the test, and blade penetration depth was set at 20 mm. Shear force was determined as the maximum force during the shearing.

Volodkevich Bite Jaws

The Volodkevich bite jaws or Volodkevich bite tenderometer (Fig. 4.5a) (Volodkevich 1938; Winkler 1939) were designed to imitate incisor teeth shearing through food sample, being used where samples can be prepared to precise cross-sectional dimensions (normally approx. 10 mm x 10 mm). The sample is rested between the 'jaws' of the rig and the shearing is performed by an upper device which is similar in geometry to an incisor tooth. The sample is positioned on the lower 'tooth' and the result is measured as the peak force required for biting through the sample (Fig. 4.5b). The limitation of the cross-section of the sample and the sometimes cumbersome holding of the sample prior to contact with the upper jaw reduced the popularity of this method in the last decade. Typically used for meat products, the results correlate with tenderness, toughness and firmness of the sample.

There is limited research using the Volodkevich for muscle food. This device was used to improve meat eating quality. Several authors studied the effect of different treatments on pork quality (Petersen et al. 1999; Maribo et al. 1999). In these works, meat samples were wrapped in plastic bags, heated for 65 minutes at 72 °C in water bath and cooled. Samples of 1x2x5 cm were dissected with the longest side parallel to the main myofibre axis. They were sheared perpendicular to the fibre axis until 80% of penetration and the bite force was expressed in Newtons. Razminowicz et al. (2006) measured the beef texture with a Volodkevich bite tendrometer mounted on a TA-XT2

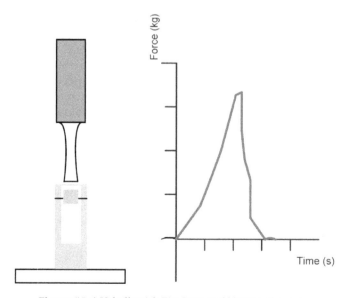

Figure 4.5 a) Volodkevich Bite Jaws and b) Typical curve.

Texture Analyzer (Stable Micro System, Surrey, UK) and Vestergaard et al. (2007) measured the quality textural quality of beef with the same shear attachment on a Karl Frank 81559 apparatus. Swain et al. (1999) used this device in an Instron Universal Testing Instrument according to Rhodes et al. (1972). The Volodkevich was also used to analyze the effect of injection of salt, tripolyphosphate and bicarbonate marinade solutions to improve the yield and tenderness of cooked pork loin (Sheard and Tali 2004). For this study, loin sections (2x1x1 cm) were cut with the muscle fibres running longitudinally. The shear force was measured using a Stevens texture analyzer (C. Stevens and Sons, Benfleet, Essex, UK) with a Volodkevitch shear cell and using a crosshead speed of 35 mm/min and a load of 20 kg. Their Volodkevich was modified to become a MIRINZ tenderometer. There, the wedges do not slide past each other as would incisor teeth but meet end to end.

Back Extrusion

Back extrusion, also called annular pumping, is derived from the penetration test. In this flow geometry, material compression induces its flow through the annulus formed between an inner cylinder and a cylindrical container. Experimental data consist in the evolution of the penetration force acting on the inner cylinder versus the displacement of the inner cylinder (or its travel time). Tests are performed under constant velocity of the inner cylinder (Perrot et al. 2011).

Meat and fish pastes and viscous mixtures (meat slurries, pastes, potted meats, terrines) must be tested in containers or in their final packaging due to their unsupported structure. In order to assess characteristics such as flow, thinning and thickening, consistency, viscosity, adhesiveness and spreadability, back extrusion test can be used.

This name was given because the food flows in the opposite direction to the plunger (compare to the compression-extrusion test in bulk analysis section). Figure 4.6 shows a schematic back extrusion test and typical curves obtained.

The back extrusion cell simulates many of the deformation patterns that occur during mastication. The flow resistance, related to the aggregate viscosity of the food specimen, is measured as the food is extruded through the annular gap between the plunger and the cell wall. The force level during extrusion is related to texture properties such as hardness and compressibility. Elasticity or recovery of the sample can be evaluated by performing multiple compressions on the same specimen.

Velazquez et al. (2007) used back extrusion analysis to assess texture in Alaska pollock surimi pastes. In their work the authors measure extrusion in a 40 mm inner diameter back extrusion ring to measure the force required

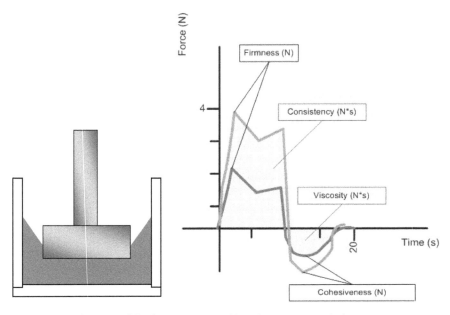

Figure 4.6 a) Back extrusion test; b) Back extrusion typical curves.
Color image of this figure appears in the color plate section at the end of the book.

for the fish paste to be extruded around a 35 mm piston disc. Samples of 30 g were introduced into the cell and extruded at 1 mm/s to 80% of its initial height. With the graphs obtained, the authors interpreted the maximum force as firmness and the area under the curve as consistency of the fish paste.

On a protein preparation obtained from washed mechanically recovered poultry meat (MRPM), Stangierski et al. (2008) evaluated texture changes using the back extrusion test. Test conditions adopted by authors were: pre-test speed 1.5 mm/s, test speed 2 mm/s, post-test speed 2 mm/s, distance 30 mm, trigger force 5 g. The maximum force was taken as a measure of firmness (N) and the area under the curve as a measure of consistency (N s). Cohesiveness (N) was interpreted as the maximum negative force. The negative area of the curve was interpreted as the index of viscosity (N s). The authors compared protein preparations of MRPM with or without transglutaminase (TG) addition. The analysis of textural parameters showed an increase in firmness, consistency, cohesiveness and index of viscosity in the preparations made with TG with time.

In order to characterize texture properties and analyze the dynamics of interactions between protein and water molecules in a myofibrillar preparation of poultry meat, Stangierski and Baranowska (2008) used the back extrusion test. Texture was tested on myofibril isolate samples, both with and without the addition of an enzymatic preparation, pre-incubated for 0.5, 1.5, 3, 4.5 and 6 h. Based on the conducted tests, the following parameters were determined: firmness (N), consistency (N*s), cohesiveness (N) and the index of viscosity (N*s). The authors concluded that transglutaminase added to the preparation catalyzed the formation of cross-linkages between peptides and aminoacids, causing significant changes in rheological properties of proteins. At the same time it enhanced the association of water molecules by myofibril proteins.

Back extrusion force is an important parameter when fish or meat patties have to be pumped through pipelines and patty-forming equipment. The data on back extrusion force may be useful for designing equipment and selection of pumps in the meat-processing industry. Effects of addition of liquid whole egg, fat and textured soy protein on textural and cooking properties of goat meat patties were studied (Singh et al. 2002). In this case, goat patty mix (50 g) was placed in a back extrusion cell having an inner diameter of 4.4 cm and a height of 10 cm. A cylindrical plunger having a diameter of 3.3 cm was forced into the cylinder to a depth of 9.5 cm at a speed of 100 mm/min. The peak force was reported as back extrusion force (N). Results showed an increase in the back extrusion force due to textured soy protein (TPS) addition and a lowered back extrusion force upon addition of both fat and egg. The increase in back extrusion force with increase in TSP level was attributed to the increase in dry matter of the patty mix. The decrease in back extrusion force with increase in fat level was attributed to its lubricating effect between meat particles and the cylinder wall, whereas the effect of egg may be attributed to the decrease in total dry matter in the patty mix in addition to its lubricating effects (Singh et al. 2002).

Bulk Analysis

Bulk analysis is used to measure particulates in meat or fish products where one sample does not represent how the consumer handles the product. Two tests are classified as bulk analysis: Kramer multiple shear test and compression-extrusion test (Ottawa cell).

In a multiple shear test, 5- or 10-blade Kramer cell drives at a constant speed through the non-uniform sample, compressing, shearing and extruding it through the slotted base. The test allows correlating texture properties with the sensory panel "mouth feel" (Kramer and Twigg 1959).

The Kramer shear cell (Fig. 4.7a) is a multi-bladed device. This kind of test allows obtaining an average of the forces required to cut through the sample of products such as tinned hams, meat pieces and shaped fish

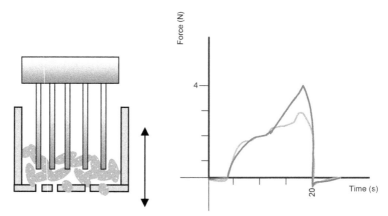

Figure 4.7 a) 5-Blade Kramer Cell Test, b) Typical curves.

Color image of this figure appears in the color plate section at the end of the book.

or poultry products (e.g., nuggets) with a variable geometry. Figure 4.7b shows typical texture analysis curves comparing a multiple shearing test on two samples of canned ham.

Boneless ham muscles (Nuñez et al. 2009) were injected (20% w/w) with a curing brine containing no plum ingredient (control), fresh plum juice concentrate (FP), dried plum juice concentrate (DP), or spray-dried plum powder (PP) at 2.5% or 5%. A standardized sample (5 cm^2) was cut, weighed and placed flat in a 10-blade Kramer shear cell. Authors compared physical, chemical and sensory properties during storage for each of the treatments. In texture properties they observed that the shear force values increased as the level of plum ingredient increased ($p < 0.05$) from 2.5% to 5%, and the highest shear values were observed in hams containing 5% FP.

Low-fat, reduced-sodium meat sausages formulated with alternative salts and gellan gum were evaluated by Totosaus and Perez-Chabela (2009). Authors compressed in a 10-blade Kramer shear press sausage samples cut into 8 cm lengths, at a constant crosshead speed of 1 mm/s. Shear-compression parameters were obtained from the deformation–compression curves, where four distinct regions may be identified the: initial slope (the blades start to compress the sample), compression slope (constant compression without shear), changing slope (both compression and shear result in the sample extrusion through the cell floor), and maximum force (sample disruption; compression, shear and extrusion occur simultaneously). The maximum force per gram and the total area (work required to compress–extrude the sample through the cell floor) were also calculated. The Kramer shear test can explain the main differences between formulations; addition of $MgCl_2$

produced a relatively large meat batter pore size, while addition of $CaCl_2$ produced a much finer pore structure.

Aussanasuwannakul et al. (2012) measured texture of the rainbow trout fillets using a 5-blade Kramer shear mounted on a texture analyzer (Texture Technologies Corp., Scarsdale, N.Y., U.S.A.), which was equipped with a 50-kg load and ran at a crosshead speed of 127 mm/min. Shear force was applied perpendicularly to the muscle fiber orientation. Force-deformation graphs were recorded and the maximum shear force (gram per gram sample) was determined. In this experience the authors studied the changes in fillet texture and compared fillets from fertile and sterile female fish.

Fillets of chicken breast were studied to investigate the effect of single-cycle and multiple-cycle high pressure (HP) treatments on the colour and texture (Del Olmo et al. 2010). Kramer cell was used for texture analysis, coupled to an Instron compression tester model 4301 (Instron, High Wycombe, Bucks, UK) working at a cross-head speed of 100 mm/min. Shear force (N/g), maximum slope (N/mm), and total energy (J/g) were determined for chicken breast fillets. Analyses were carried out at room temperature (21–23 ºC) on four samples (50×15×10 mm) of chicken breast fillets per treatment, in each of the two HP experiments. The authors reflected that texture parameters increased with the length of treatment, but beyond a certain severity of treatment these parameters declined.

Chicken nuggets and other poultry products are coated by a number of layers to provide flavour and texture as well as to add value. Factors such as cooking method, duration of cooking and meat choice extend the range of potential texture-affecting concerns. The Kramer shear cell allows the testing of a larger sample size and therefore can have the advantage of testing using an averaging effect, which is advantageous for the repeatable testing of highly non-uniform products. Varela et al. in 2008 described a new method to assess the texture of crispy-crusted foods with a soft, high-moisture core. This method, applied to commercial pre-cooked chicken nuggets, combines characteristics derived from the force/displacement curves of the whole sample with characteristics of the simultaneously emitted sound. The use of a Kramer shear cell to perform the test proved to be an effective technique for characterizing the texture of chicken nuggets after different final cooking processes. The force curves of the samples differed with the cooking process. Deep-fried samples and those cooked in a conventional oven presented jagged force characteristic of crispy products. The curve profiles of microwaved samples were drastically different and typical of tough, gummy products. It was found that although the moisture and fat contents of the core and crust are closely related to the texture characteristics, samples with similar water contents can have very dissimilar crispness characteristics. The fat content of the core did not change significantly with the final cooking process in any of the samples.

96   *Methods in Food Analysis*

Bulk compression is when the sample is compressed in three dimensions. It is commonly used for meat or fish products that may contain particulates such as onions, coarse muscle fibers and nuts, where a chosen number of pieces or weight of sample is tested at the same time. The most common compression-extrusion test was developed by Voisey (1971). Figure 4.8 show a typical Ottawa cell developed for this analysis. The maximum force peak describes firmness.

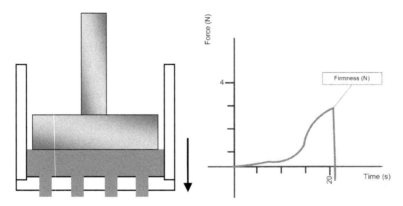

Figure 4.8 a) Ottawa cell; b) Measurement of firmness using the Ottawa cell.

4.3.2 Compression Test

Compression testing is a common test for meat or fish products. A sample is placed on a flat surface and a compression plate is lowered onto the sample. Force, position or percentage of the original height of the sample may be used to interpret the compression test. In a single compression test it is possible to evaluate the hardness or firmness, and fracturability of the sample (Friedman et al. 1963).

A special derivative of the compression test is the texture profile analysis (TPA, Fig. 4.9). TPA has 4 steps: first compression, relaxation, second compression and a final relaxation. Texture attributes such as hardness, cohesiveness, springiness, chewiness, resilience, gumminess, adhesiveness or stickiness may be measured with a TPA test (Fig. 4.10). Several authors doubt if the second compression offers additional information, because a high degree of compression fractures the sample after the first compression and the structures are different during the two compressions. Meullenet and Gross (1999) evaluated 24 food samples (meat, vegetables, confectionery, and snacks) with a sensory panel and a modified instrumental TPA, where the samples were compressed to 80% of the initial height. They did not find

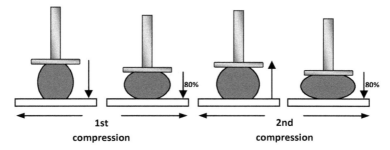

Figure 4.9 Typical TPA test.

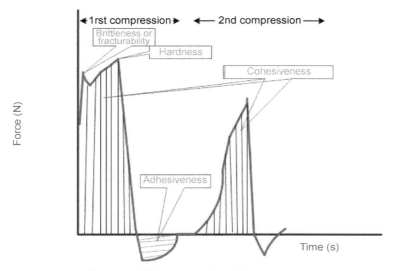

Figure 4.10 TPA curves and typical parameters.

Color image of this figure appears in the color plate section at the end of the book.

that the double compression test offered any significant improvements over the single compression test. This conclusion supports the above opinion.

Textural and protein characteristics of crisp grass carp (CGC) and grass carp (GC) fillets were studied using TPA analysis. The authors observed changes in fish proteins correlated to the differences in TPA texture characteristics of the fillets. The higher TPA parameters of CGC are related to a higher content of myofibrillar, sarcoplasmic, stromal proteins, sulfur aminoacids and hydrophobic aminoacids. For texture analyses, fillets of 6 cm^3 (2 cm x 2 cm x 1.5 cm) of the dorsal fish part were cut. The samples were examined using a 35-mm diameter cylindrical probe. The test conditions involved two consecutive cycles of 30% compression with

5 s between cycles. The pretest speed, test speed and posttest speed were 1, 1 and 5 mm/s, respectively. The measured parameters were hardness (g), fracturability (g), springiness, cohesiveness, chewiness (g) and resilience (Lin et al. 2012).

Changes caused by gravading and freezing are likely to affect the textural properties of the muscle tissue of 1-year-old farmed rainbow trout, which were examined using TPA (Michalczyk and Surówka 2009). Samples for the TPA were collected from fresh fillets and from gravads directly after their production and freezing. Samples with the skin left on, 20 mm in diameter and 9 mm thick, were cut by a cork borer from the middle part of the fillet over the lateral line, placed with the skin underneath on an Platform and then compressed and decompressed twice with a 45 mm diameter cylindrical probe until 50% deformation was achieved. The rate of the probe was 2 mm/s and the time interval between strokes was 5 s. Hardness was defined as the final force required for 50% deformation of the sample by the probe. Cohesiveness, which represents the forces holding the product together, was calculated as the ratio of areas delimited by the curves of the second and the first bite. Springiness, defined as the rate at which a deformed sample returns to its undeformed condition after the deforming force is removed, was expressed as the ratio of the time measured between the start of the second area and the second probe reversal divided by the time measured between the start of the first area and the first probe reversal. Chewiness was calculated by multiplying hardness, cohesiveness and springiness. The textural characteristics of gravad undergo change, resulting in muscle tissue which, compared to unprocessed fish, shows greater cohesiveness and chewiness measured by TPA tests as well as lower stress decay during relaxation.

The relationship between the parameters obtained by instrumental TPA at low deformation, applied to the raw whole gilthead sea bream (*Sparus aurata*) and the sensory texture profile analysis of the cooked fish were studied by Carbonell et al. (2003). TPA was performed using a 24.5 mm diameter flat plunger and a cross-head speed of 5 mm/s. TPA measurements were carried out on the whole fish, compressing each specimen on the dorsal side after removing the lateral fin about 2 cm from the gill. The fish were compressed twice by 7 mm with no waiting time between the two compressions and the TPA parameters (hardness, cohesiveness, springiness, and chewiness) were obtained. The authors concluded that a simple nondestructive TPA test on the raw whole fish (fresh or frozen and thawed) may be used to obtain predictive information about some sensory characteristics of the final cooked product.

Mechanical properties by TPA and puncture test of fish gels were studied (Veláquez et al. 2007). This work evaluated the effect of adding soluble proteins (SP) from surimi wash water (SWW) in commercial Alaska pollock surimi grade FA. Cylindrical surimi gel samples (20.8 mm diameter and 25 mm height) were tested by TPA using a 50 mm cylindrical probe. Samples were compressed at 1 mm/s to 75% of their initial height. Hardness, fracturability, springiness, cohesiveness and chewiness values were determined. SWW-SP could be used at 10–30 g/kg to improve mechanical properties of Alaska pollock surimi gels. In this case an increase in hardness, fracturability, springiness and chewiness were detected while changes in cohesiveness were very low and no significant effect was observed.

Changes to postmortem muscle texture in white shrimp (*Litopenaeus vannamei*) with necrotized hepatopancreatitis bacteria (NHPB) were evaluated using TPA. The TPA was used to evaluate five parameters: hardness, brittleness, cohesiveness, elasticity and adhesiveness with the aim to evaluate the effects of NHPB infection on the postmortem texture of raw and cooked shrimp muscle (Ávila-Vila et al. 2012). Raw and cooked shrimp muscles were transversely cut at 1 cm between the first and the second segments. To estimate the TPA, the maximum effort required to cut the muscle (hardness) was registered; compression forces at 75 and 90% (respectively) of the original gel sample height were used to compute compression hardness, fracturability, elasticity and cohesiveness. The effect of NHPB on the textural properties of the muscle cannot be detected during the first days of the infection, but changes can be documented after 18 days.

The effects of composition (protein and fat content) and cooking temperature (heating rate) on the textural characteristics were assessed in Bologna sausages (Jiménez-Colmenero et al. 1995). Texture Profile Analysis (TPA) was applied to five Bologna cores (diam 2.5 cm, height 2 cm) per treatment that were axially compressed to 50% of their original height. Force–time deformation curves were derived with a 5 kN load cell applied at a cross-head speed of 50 mm/min. The following attributes were calculated: hardness (Hd)—peak force (N) required for first compression; cohesiveness (Ch)—ratio of active work done under the second compression curve to that done under the first compression curve (dimensionless); springiness (Sp)—distance (cm) the sample recovers after the first compression; and chewiness (Cw)— Hd x Ch x Sp (N.cm). The authors built a regression model to study the simultaneous effect of two composition variables and one processing variable. Cooking temperature was the one that least affected the TPA parameters of the meat emulsion. This variable contributed most to the regression model for hardness and chewiness through interaction with the other two variables.

An alternative to obtain low-fat formulated meats is to use functional ingredients that mimic the properties of fat in emulsified cooked meat products. These fat replacements or substitutes contribute to reduce calories to formulated meats without altering their flavor, juiciness, mouth feel, viscosity or other organoleptic and processing properties. With this objective Totoause and Perez-Chabela (2009) studied the effect of incorporation of calcium or magnesium chloride in addition to potassium chloride on functional, textural and microstructural properties in low-fat, reduced-sodium meat batters formulated with gellan gum. Sausage treatments were analyzed using a 25 kg load cell and a 25 mm diameter acrylic probe. Sausage samples were cut in 20 mm length cylinders and axially compressed 50% of their original height. Constant cross-head speed was 1 mm/s with a 5 s waiting period. Results showed gellan gum incorporation plus magnesium or calcium chloride produced lower hardness and higher cohesiveness values in the studied low-fat, reduced-sodium meat batters compared to the control; however, all other Texture Profile Analysis (springiness, gumminess and resilience) parameters were higher than the control.

Cilla et al. (2006) studied the effects of long-term storage on the quality and acceptability of two common ham commercial formats: refrigerated vacuum-packaged boneless whole dry-cured hams and frozen vacuum-packaged ham cuts. Samples (1 cm in diameter and 1.5 cm high) of both semimembranosus muscle (SM) and biceps femoris (BF) muscles were compressed axially in two consecutive cycles of 50% compression with a flat plunger 50 mm in diameter. The cross-head moved at a constant speed of 1 mm/s with 5 s between cycles. The following texture parameters were measured from the force-deformation curves: hardness, springiness, cohesiveness, adhesiveness and chewiness. Significant changes involved loss of odour and flavour, increased adhesiveness and modification of hardness, the SM became more tender while BF became harder, leading to a higher textural homogeneity.

The same authors (Cilla et al. 2005) had studied the evolution of biochemical and instrumental texture parameters, and others attributes of 12-month dry-cured hams maintained up to 26 months under "bodega" conditions (18 °C, 75% relative humidity), in order to investigate the influence of extended ripening on their sensory profile and consumer acceptability. The same TPA conditions test (50% compression) was used. The same textural changes were observed in SM and BF muscle, particularly regarding hardness, which increased significantly until the 20th month, and decreased significantly after that. Evolution of adhesiveness had an opposite behavior, since it decreased until the 20th month and increased significantly ($p < 0.05$) at the end of storage.

Slices of dry-cured hams were stored during 8 weeks under vacuum and two different modified atmospheres, in order to study the effects on texture and other characteristics during that period (García-Esteban et al. 2004). Compression was performed with a cylindric probe of one inch diameter. Two compression cycles were carried out until 30% of deformation, applying a cross-head speed of 3 mm/s and a constant speed of 1 mm/s. The results revealed that modified atmosphere packaging preserved samples from hardening and deterioration of textural properties more efficiently than vacuum packaging. In summary, it could be concluded that, with regard to colour, texture, moisture and microbiological stability, dry-cured ham stored in vacuum and modified atmospheres did not show clear differences.

Texture profile analysis (TPA) was used to evaluate sausage texture by Santos et al. (2004). Samples (2x1.5 cm) were compressed twice to 50% of their original height using a cylindrical probe P/25. The objective of the authors was to evaluate the possibility of using pork legs enriched in $n-3$ fatty acids and α-tocopherol in the manufacture of pork sausages. Textural features were not affected by the cooked ham enrichment in $n-3$ PUFAs, since no significant differences in the results of the TPA among different experimental cooked hams were found.

4.3.3 Penetration Test

Penetration test has been used to analyze gels, pates, and meat or fish fillets. In meat pates, one can detect differences between two formulations with fat substitutes. Texture parameters in low-salt fish products made with transglutaminase as binding agent may be studied using a puncture test.

To study the effect of fat content on the physicochemical and nutritional characteristics and oxidation stability of pork liver pates product, Estevez et al. (2005) used penetration test on different formulation of this product. The penetration test was performed with a 10 mm diameter cylinder probe which penetrated a depth of 8 mm within the sample. Force in compression was recorded at a crosshead speed of 1.5 mm/s. The maximum force evaluated hardness (N) of the sample. In agreement with others works, the authors found that high fat content pates presented significantly lower values of hardness than low fat content pates ($p<0.05$). The effect of fat on texture results in less firm and more juicy products.

The effect of a preheating treatment on mechanical properties of beef gels obtained with transglutaminase was determined using the puncture test. The test was performed by compressing samples to 75% of the initial height using a compression speed of 1 mm/s with a 12 mm spherical probe. The samples were penetrated at the center and breaking force (N), deformation (cm), and penetration work (N.cm) were calculated.

To determine the effects of using dairy proteins and microbial transglutaminase as binding agents in low-salted restructured silver carp (*Hypophthalmichthys molitrix*) products, penetration test was performed (Uresti et al. 2004). Cylindrical samples (1.87 cm diameter and 3 cm length) were obtained. The puncture test was performed, compressing samples to 75% of the initial height using a compression speed of 1 mm/s and a cylindric probe of 1.2 cm diameter. The breaking force (g), deformation (cm) and work of penetration (g.cm) were measured. The results obtained showed that MTGase required the addition of salt to fish paste in order to improve the mechanical properties. The use of dairy proteins and MTGase had an improving effect on the mechanical properties of low-salt products. To determine the effect of various durations of electrical stimulation of salmon muscle on fillet quality Roth et al. (2010) used the puncture test. In this work, the puncture test was assessed in 2 locations directly on the salmon fillets (skin on), transverse to the muscle fiber orientation. The probe (20 mm of diameter) was programmed to penetrate 80% into the initial fillet height and maximum forces were recorded in addition to forces at 60% compression. Using this analysis, the authors conclude that a 6 s current duration is favorable for stunning fish as pre-rigor times and other quality attributes are slightly changed.

4.3.4 Other Texture Methods

Tension

A method developed in meat research is the tensile strength evaluation. Strength analysis may be done using the tensile grips as shown in Fig. 4.11. With the aim of testing cooked meat sausages, Herrero et al. (2008) determined their tensile parameters (breaking strength and energy to fracture) and correlated these results to TPA parameters and physico-chemical characteristics. In this work textural properties of some cooked meat sausages (chopped, mortadella and galantines) were studied. Pieces

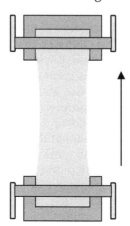

Figure 4.11 Tensile Grips.

were cut in a dumbbell shape, approximately 7.5 cm × 2 cm × 0.2 cm thickness. For analysis, one tensile grip was fixed to the base of the textural analyser, while the other one was attached to the load cell (5 kg). Initial grip separation was 12.5 mm and cross-head speed was 1.0 mm/s until rupture. Rupture force was taken as the maximum force peak height (N) required for breaking the sample. Breaking strength (N/cm^2) was obtained by dividing the rupture force by the cross-sectional area (thickness × width) of the portions. Energy to fracture (N*mm) was calculated as the area under the deformation curve. The determination of the breaking strength (BS) and the energy to fracture (EF) by tensile test may be used combined with the TPA, to determine textural properties of cooked meat sausages.

"Finger" Method

This is a non-destructive method where in a cylindric aluminium probe with a ball tip is used to simulate compression with a human finger (Fig. 4.12). The compression test is performed by taking into consideration the fillet thickness when depth is selected. When the test is running, the ball probe is allowed to press the muscle fiber to a selected depth at a speed of 1 mm/s and then it returns to the starting position. Using this "finger method" to evaluate the firmness in salmon (*Salmo salar*) fillet, Sigurgisladottir et al. (1997) found that no difference in force was observed on the head and middle parts when a 5 mm compression was made by a ball probe on the fillet samples of natural thickness.

In another work, Sigurgisladottir et al. (1999) applied three methods (flat cylinder, ball probe and shear blade) to evaluate the quality of salmon fillets. Spherical probe was applied on both fillets with natural thickness and constant-thickness samples. 5 mm depth was selected as a maximum travel

Figure 4.12 Ball probe for "finger" method.

distance without breaking muscular fibers. The results showed that the ball probe affected the muscle less than a flat cylinder, and could compress the sample more without damaging the muscle structure.

In Japanese research the spherical probe seems to be preferred in measurements of fish, especially sushi.

4.4 Conclusion

Meat and fish are complex biological materials, which are heterogeneous and anisotropic. Knowledge of textural properties in meat and fish products is crucial for success in new developments. Viscosity, thickness, hardness, firmness, chewiness, toughness, stringiness, and brittleness (fracturability) are some of the basic characteristics that are associated with the physical properties of meat and fish products that may have the greatest influence on consumer acceptance. The validity and repeatability of these type of measurements depend on both instrumentation and methodology.

References

Ahnström, M.L., Enfält, A.C., Hansson, I. and Lundström, K. 2006. Pelvic suspension improves quality characteristics in M. semimembranosus from Swedish dual purpose young bulls. Meat Science. 72: 555–559.

Ahnström, M.L., Hunt, M.C. and Lundström, K. 2012. Effects of pelvic suspension of beef carcasses on quality and physical traits of five muscles from four gender–age groups. Meat Science. 90: 528–535.

Alba De, M., Montiel, R., Bravo, D., Gaya, P. and Medina, M. 2012. High pressure treatments on the inactivation of *Salmonella Enteritidis* and the physicochemical, rheological and color characteristics of sliced vacuum-packaged dry-cured ham. Meat Science. 91: 173–178.

AMSA. 1995. Research Guidelines for Cookery, Sensory Evaluation, and Instrumental Tenderness Measurements of Fresh Meat. American Meat Science Association, Chicago, IL.

Ando, M., Nishiyabu, A., Tsukamasa, Y. and Makinodan, Y. 1999. Postmortem softening of fish muscle during chilled storage as affected by bleeding. Journal of Food Science. 64(3): 423–428.

Aussanasuwannakul, A., Weber, G.M., Salem, M., Yao, J., Slider, S., Manor, M.L. and Kenney, P.B. 2012. Effect of Sexual Maturation on Thermal Stability, Viscoelastic Properties, and Texture of Female Rainbow Trout, *Oncorhynchus mykiss*, Fillets. Journal of Food Science. 71(1): 77–83.

Avila-Villa, L.A., Garcia-Sanchez, G., Gollas-Galvan, T., Hernandez-Lopez, J., Lugo-Sanchez, M.E., Martinez-Porchas, M. and Pacheco-Aguilar, R. 2012. Textural changes of raw and cooked muscle of shrimp (*Litopenaeus vannamei*), infected with necrotizing hepatopancreatitis bacterium (NHPB). Journal of Texture Studies. 43: 453–458.

Barbut, S. 2013. Frying—Effect of coating on crust microstructure, color, and texture of lean meat portions. Meat Science. 93: 269–274.

Barroso, M., Careche, M. and Borderias, A.J. 1998. Quality control of frozen fish using rheological techniques. Trends in Food Science and Technology. 9: 223–229.

Baublits, R.T., Meullenet, J.F., Sawyer, J.T., Mehaffey, J.M. and Saha, A. 2006. Pump rate and cooked temperature effects on pork loin instrumental, sensory descriptive and consumer-rated characteristics. Meat Science. 72: 741–750.

Beltrán-Lugo, A., Maeda-Martínez, A.N., Pacheco-Aguilar, R., Nolasco-Soria, H.G. and Ocaño-Higuera, V.M. 2005. Physical, textural, and microstructural properties of restructured adductor muscles of 2 scallop species using 2 cold-binding systems. Journal of Food Science. 70(2): E78–E84.

Bouton, P.E. and Harris, P.V. 1978. Factors affecting tensile and Warner-Bratzler shear values of raw and cooked meat. Journal of Texture Studies. 9(4): 395–413.

Brady, P.L. and Hunecke, M.E. 1985. Correlations of sensory and instrumental evaluation of roast beef texture. Journal of Food Science. 50: 300–303.

Bratzler, L.J. 1932. Measuring the tenderness of meat by mechanical shear. M.Sc. Thesis, Kansas State University, Manhattan, U.S.A.

Bratzler, L.J. 1949. Determining the tenderness of meat by use of the Warner-Bratzler method. Proc. 2nd Reciprocal Meat Conference, pp. 117–121.

Cáceres, E., García, M.L. and Selgas, M.D. 2006. Design of a new cooked meat sausage enriched with calcium. Meat Science. 73: 368–377.

Cáceres, E., García, M.L., Toro, J. and Selagas, M.D. 2004. The effect of fructooligosaccharides on the sensory characteristics of cooked sausages. Meat Science. 68: 87–96.

Carbonell, I., Duran, L., Izquierdo, L. and Costell, E. 2003. Texture of cultured gilthead sea bream (*Sparus aurata*): instrumental and sensory measurement. Journal of Texture Studies. 34: 203–217.

Castro-Briones, M., Calderón, G.N., Velazquez, G., Rubio, M.S., Vázquez M. and Ramírez, J.A. 2009. Mechanical and functional properties of beef products obtained using microbial transglutaminase with treatments of pre-heating followed by cold binding. Meat Science. 83: 229–238.

Cavitt, L.C., Meullenet, J.F., Gandhapuneni, R.K., Youm, G.W. and Owens, C.M. 2005a. Rigor development and meat quality of large and small broilers and the use of Allo-Kramer Shear, Needle Puncture, and Razor Blade Shear to measure texture. Poult. Sci. 84: 113–118.

Cavitt, L.C., Meullenet, J.F., Xiong, R. and Owens, C.M. 2005b. The relationship of razor blade shear, Allo-Kramer shear, Warner-Bratzler shear and sensory tests to changes in tenderness of broiler breast fillets. J. Muscle Foods 16: 223–242.

Cavitt, L.C., Owens, C.M., Meullenet, J.-F., Gandhapuneni, R.K. and Youm, G.W. 2001. Rigor development and meat quality of large and small broilers and the use of Allo-Kramer shear, needle puncture, and razor blade shear to measure texture. Poultry Sci. 80(Suppl. 1): 113–118.

Cavitt, L.C., Youm, G.W., Meullenet, J.F., Owens, C.M. and Xiong, R. 2004. Prediction of poultry meat tenderness using razor blade shear, Allo-Kramer shear, and sarcomere length. Journal of Food Science. 69(1): SNQ11–SNQ15.

Chamberlain, A.I., Kow, F. and Balasubramaniam, E. 1993. Instrumental method for measuring texture in fish. Food Aust. 45: 439–443.

Choudhury, G.S., Gogoi, B.K. and Oswalt, A.J. 1998. Twin-screw extrusion pink salmon muscle and rice flour blends: effects of kneading elements. Journal of Aquatic Food Product Technology. 7: 60–91.

Christensen, M., Andersen, E., Christensen, L., Andersen, M.L. and Baron, C.L. 2011. Textural and biochemical changes during ripening of old-fashioned salted herrings. Journal of the Science of Food and Agriculture. 91: 330–336.

Chung, S.L. and Merritt, J.H. 1991. Physical measures of sensory texture in thawed sea scallop meat. International Journal of Food Science and Technology. 26: 207–210.

Cilla, I., Martínez, L., Beltrán, J.A. and Roncalés, P. 2005. Factors affecting acceptability of dry-cured ham throughout extended maturation under "bodega" conditions. Meat Science. 69(4): 789–795.

Cilla, I., Martínez, L., Beltrán, J.A. and Roncalés, P. 2006. Effect of low-temperature preservation on the quality of vacuum-packaged dry-cured ham: Refrigerated boneless ham and frozen ham cuts. Meat science. 7(1): 12–21.

Combes, S., Postollec, G., Cauquil, L. and Gidenne, T. 2010. Influence of cage or pen housing on carcass traits and meat quality of rabbit. Animal. 4(2): 295–302.

Cover, S., Hostetler, R.L. and Ritchey, S.J. 1962. Tenderness of beef. IV. Relations of shear force and fiber extensibility to juiciness and six components of tenderness. Journal of Food Science. 27: 527–536.

De Alba, M., Montiel, R., Bravo, D., Gaya, P. and Medina, M. 2012. High pressure treatments on the inactivation of *Salmonella Enteritidis* and the physicochemical, rheological and color characteristics of sliced vacuum-packaged dry-cured ham. Meat Science. 91: 173–178.

Del Olmo, A., Morales, P., Ávila, M., Calzada, J. and Nuñez, M. 2010. Effect of single-cycle and multiple-cycle high-pressure treatments on the colour and texture of chicken breast fillets. Innovative Food Science and Emerging Technologies. 11: 441–444.

Estevez, M., Ventanas, S. and Cava, R. 2005. Physicochemical properties and oxidative stability of liver pate as affected by fat content. Food Chemistry. 92: 449–457.

Fernández-Segovia, I., Camacho, M.M., Martínez-Navarrete, N., Escriche, I. and Chiralt, A. 2003. Structure and colour changes due to thermal treatments in desalted cod. Journal of Food Processing and Preservation. 27: 465–474.

Friedman, M.M., Whitney, J.E. and Szczesniak, A.S. 1963. The texturometer: a new instrument for objective texture measurement. Journal of Food Science. 28: 390–396.

FTC, Food Technology Corporation. Test cell and fixtures. http://www.foodtechcorp.com/documents/brochures/test_cells_fixtures/. Last online access December 9, 2012.

García-Esteban, M., Ansorena, D. and Astiasar, I. 2004. Comparison of modified atmosphere packaging and vacuum packaging for long period storage of dry-cured ham: effects on colour, texture and microbiological quality. Meat science. 67(1): 57–63.

García-Segovia, P., Andrés-Bello, A. and Martínez-Monzó, J. 2011. Chapter 9: Texture and Microstructure. *In*: Rui M.S. Cruz (ed.). Practical Food and Research. Nova Science Publishers, Hauppauge NY, USA.

Gautam, A., Choudhury, G.S. and Gogoi, B.K. 1997. Twin screw extrusion of pink salmon muscle: effect of mixing elements and feed composition. Journal of Muscle Foods. 8: 265–285.

Girard, I., Aalhus, J. L., Basarab, J. A., Larsen, I. L. and Bruce, H. L. 2011. Modification of muscle inherent properties through age at slaughter, growth promotants and breed crosses. Canadian Journal of Animal Science, 91: 635–648.

Girard, I., Aalhus, J.L., Basarab, J.A., Larsen, I.L. and Bruce, H.L. 2012. Modification of beef quality through steer age at slaughter, breed cross and growth promotants. Canadian Journal of Animal Science. 92: 175–188.

Girard, I., Bruce, H.L., Basarab, J.A., Larsen, I.L. and Aalhus, J.L. 2012. Contribution of myofibrillar and connective tissue components to the Warner–Bratzler shear force of cooked beef. Meat Science. 92: 775–782.

Guerrero, L., Gou, P. and Arnau, J. 1999. The influence of meat pH on mechanical and sensory textural properties of dry-cured ham. Meat Science. 52: 267–273.

Harper, B.A., Barbut, S., Lim, L.T. and Marcone, M.F. 2012. Microstructural and textural investigation of various manufactured collagen sausage casings. Food Research International. 49: 494–500.

Henrickson, R.L., Marsden, J.L., and Morrison, R.D. 1974. An evaluation of a method for measuring shear force for an individual muscle fiber. Journal of Food Science. 39: 15–20.

Herrero, A.M., de la Hoz, L., Ordóñez, J.A., Herranz, B., Romero de Ávila, M.D. and Cambero, M.I. 2008. Tensile properties of cooked meat sausages and their correlation with texture profile analysis (TPA) parameters and physico-chemical characteristics. Meat Science. 80(3): 690–696.

Honikel, K.O. 1998. Reference methods for the assessment of physical characteristics of meat. Meat Science. 49(4): 447–457.
Houben, J.H., Bakker, W.A.M. and Keizer, G. 2005. Effect of trisodium phosphate on slip and textural properties of hog and sheep natural casings. Meat Science. 69: 209–214.
Iseya, Z., Sugiura, S. and Saeki, H. 1996. Procedure for mechanical assessment of textural change in dried fish meat. Fisheries Science. 62: 772–775.
Jiménez-Colmenero, F., Barreto, G., Mota, N. and Carballo, J. 1995. Infl uence of protein and fat content and cooking temperature on texture and sensory evaluation of bologna sausage. LWT - Food Science and Technology. 28: 481–487.
Kanoh, S., Polo, J.M.A., Kariya, Y., Kaneko, T., Watabe, S. and Hashimoto, K. 1988. Heat induced textural and histological changes or ordinary and dark muscles of yellowfin tuna. Journal of Food Science. 53: 673–678.
Kramer, A. and Twigg, B.A. 1959. Principles and instrumentation for the physical measurement of food quality with special reference to fruit and vegetable products. Adv. Food Res. 9: 153–220.
Lee, Y.S., Owens, C.M., and Meullenet, J.F. 2008. The meullent-owens razon shear (MORS) for predicting poultry meat tenderness: its applications and optimization. Journal of Texture Studies. 39: 655–672.
Lin, W.L., Zeng, Q.X., Zhu, Z.W. and Song, G.S. 2012. Relation between protein characteristics and TPA texture characteristics of Crisp Grass Carp (*Ctenopharyngodon idellus c. et v*) and Grass Carp (*Ctenopharyngodon idellus*). Journal of Texture Studies. 43: 1–11.
Lyon, B.G. and Lyon, C.E. 1998. Assessment of three devices in shear tests of cooked breast meat. Poultry Science. 77: 1585–1590.
Maribo, H., Ertbjerg, P., Andersson, M., Barton-Gade, P. and Moller, A.J. 1999. Electrical stimulation of pigs- effect on pH fall, meat quality and Cathepsin B+L activity. Meat Science. 52: 179–187.
Melito, H.S. and Daubert, C.R. 2011. Rheological Innovations for Characterizing Food Material Properties. Annual Review of Food Science and Technology. 2: 153–179.
Meullenet, J.F. and Gross, J. 1999. Instrumental single and double compression tests to predict sensory texture characteristics of foods. Journal of Texture Studies. 30(2): 167–180.
Meullenet, J.F., Jonville, E., Grezes, D. and Owens, C.M. 2004. Prediction of the texture of cooked poultry pectoralis major muscles by near-infrared reflectance analysis of raw meat. J. Texture Studies. 35: 573–585.
Michalczyk, M. and Surowka, K. 2009. Microstructure and instrumentally measured textural changes of rainbow trout (*Oncorhynchus mykiss*) gravads during production and storage. J. Sci. Food Agric. 89: 1942–1949.
Moller, A.J. 1980. Analysis of Warner-Bratzler shear pattern with regard to myofibrillar and connective tissue components of tenderness. Meat Science. 5: 247–260.
Möller, A.J., Sorensen, S.E., Larsen, M. 1981. Differentiation of myofribrillar and connective tissue strength in beef muscles by Warner-Bratzler shear parameters. Journal of Texture Studies. 12(1): 71–83.
Montero, P. and Borderias, J. 1990. Behaviour of myofibrillar proteins and collagen in hake (*Merluccius merluccius* L.) muscle during frozen storage and its effects on texture. Zeitschrift für Lebensmittel-Untersuchung und Forschung. 190: 112–117.
Montero, P. and Borderias, J. 1992. Influence of myofibrillar proteins and collagen aggregation on the texture of frozen hake muscle. In Quality Assurance in the Fish Industry. Huss, H.H., Jakobsen, M. and Liston, J. (eds.). pp. 149–167, Elsevier Science Publisher BV, Amsterdam.
Morkore, T. and Einen, O. 2003. Relating sensory and instrumental texture analyses of Atlantic salmon. Journal of Food Science. 68: 1492–1497.
Nuñez de Gonzaleza, M.T., Hafleyb, B.S., Bolemanc, R.M., Millerd, R.M., Rheed, K.S. and Keeton, T.J. 2009. Qualitative effects of fresh and dried plum ingredients on vacuum-packaged, sliced hams. Meat Science. 83(1): 74–81.

Oeckel, M.J. van, Warnants, N. and Boucque, Ch. V. 1999. Pork tenderness estimation by taste panel, Warner-Bratzler shear force and on-line methods. Meat Science. 53: 259–267.
Pérez-Won, M., Barraza, M., Cortés, F., Madrid, D., Cortés, P., Roco, T., Osorio, F. and Tabilo-Munizaga, G. 2006. Textural characteristics of frozen blue squat lobster (*Cervimunida Johni*) tails of measured by instrumental and sensory methods. Journal of Food Process Engineering. 29(5): 519–531.
Perrot, A., Mélinge, Y., Estellé, P., Rangeard, D. and Lanos. C. 2011. The back extrusion test as a technique for determining the rheological and tribological behaviour of yield stress fluids at low shear rates. Applied Rheology. 21: 53642.
Petersen, J.S., Henckel, P., Maribo, H. and Oksbjerg, N. 1997. Muscle metabolic traits, post mortem-pH-decline and meat quality in pigs subjected to regular physical training and spontaneous activity. Meat Science. 46(3): 259–215.
Razminowicz, R.H., Kreuzer, M. and Scheeder, M.R. 2006. Quality of retail beef from two grass-based production systems in comparison with conventional beef. Meat Science. 73: 351–361.
Rees, M.P., Trout, G.R., Robyn D. and Warner, R.D. 2002. Effect of calcium infusion on tenderness and ageing rate of pork *m. longissimus thoracis* et *lumborum* after accelerated boning. Meat Science. 61: 169–179.
Rhodes, D.H., Jones, C.D., Chrystall, B.B. and Harries, J.M. 1972. The relationship between subjective assessments and a compressive test on roast beef. J. Texture Studies 3: 298–309.
Roth, B., Nortvedt, R., Slinde, E., Foss, A., Grimsbø, E. and Stien, L.H. 2010. Electrical stimulation of Atlantic salmon muscle and the effect on flesh quality. Aquaculture. 301: 85–90.
Santos, C., Ordóñez, J.A., Cambero, I., D'Arrigo, M. and Hoza, L. 2004. Physicochemical characteristics of an alpha-linolenic acid and alpha-tocopherol-enriched cooked ham. Food Chemistry. 88(1): 123–128.
Sawyer, J.T., Baublits, R.T., Apple, J.K., Meullenet, J.F., Johnson, Z.B. and Alpers, T.K. 2007. Lateral and longitudinal characterization of color stability, instrumental tenderness, and sensory characteristics in the beef semimembranosus. Meat Science. 75: 575–584.
Schubring, R. 2000. Instrumental and sensory evaluation of the texture of fish fingers. Deutsche Lebensmittel-Fundschau. 96(2): 210–221.
Sheard, P.R. and Tali, A. 2004. Injection of salt, tripolyphosphate and bicarbonate marinade solutions to improve the yield and tenderness of cooked pork loin. Meat Science. 68: 305–311.
Sigurgisladottir, S., Hafsteinsson, H., Jonsson, A., Lie, O., Nortvedt, R., Thomassen, M. and Torrissen, O. 1999. Textural properties of raw salmon fillets as related to sampling method. J. Food Sci. 64: 99–104.
Sigurgisladottir, S., Torrissen, O., Lie, O., Thomassen, M. and Hafsteinsson, H. 1997. Salmon quality: Methods to determine the quality parameters. Rev. Fish. Sci. 5: 223–252.
Singh Gujral, H., Amrit Kaur, A., Singh, N. and Singh Sodhi, N. 2002. Effect of liquid whole egg, fat and textural soy protein on the textural and cooking properties of raw and baked patties from goat meat. Journal of Food Engineering. 53: 377–385.
Stangierski, J. and Baranowska, H.M. 2008. Analysis of texture and dynamics of water binding in enzymatically modified myofibrillar preparation obtained from washed mechanically recovered poultry meat. Eur. Food Res. Technol. 226: 857–860.
Stangierski, J., Baranowska, H.M., Rezler, R. and Kijowski, K. 2008. Enzymatic modification of protein preparation obtained from water-washed mechanically recovered poultry meat. Food Hydrocolloids. 22(8): 1629–1636.
Steffe, J.F. 1996. Introduction to Rheology. *In*: Rheological Methods in Food Process Engineering. 2nd Ed. Freeman press. Michigan, USA.
Swain, M.V.L., Gigiel, A.J. and Limpens, G. 1999. Textural effect of chilling hot Longissimus dorsi muscle with solid CO2. Meat Science. 51: 363–370.
Szczesniak, A.S. 2002. Texture is a sensory property. Food Quality and Preference. 13: 215–225.

Totosaus, A. and Perez-Chabela, M.L. 2009. Textural properties and microstructure of low-fat and sodium-reduced meat batters formulated with gellan gum and dicationic salts. LWT-Food Science and Technology. 42: 563–569.
Tseng, Y.C., Xiong, Y.L., Webster, C.D., Thompson, K.R. and Muzinic, L.A. 2005. Quality changes in Australian red claw crayfish (*Cherax quadricarinatus*) stored at 0°C. Journal of Applied Aquaculture. 12(4): 53–66.
Uresti, R., Ramirez, J., Lopez, N. and Vazquez M. 2003. Negative effect of combining microbial transglutaminase with low methoxyl pectins on the mechanical properties and colour attributes of fish gels. Food Chemistry. 80(4): 551–556.
Uresti, R., Tellez-Luis, S.J., Ramirez, J. and Vazquez M. 2004. Use of dairy proteins and microbial transglutaminase to obtain low-salt fish products from filleting waste from silver carp (*Hypophthalmichthys molitrix*). Food Chemistry. 86: 254–262.
Varela, P., Salvador, A. and Fiszman, S. 2008. Methodological developments in crispness assessment: Effects of cooking method on the crispness of crusted foods. LWT-Food Science and Technology. 41(7): 1252–1259.
Veland, J.O. and Torrissen, O.J. 1999. The texture of Atlantic salmon (*Salmo salar*) muscle as measured instrumentally using TPA and Warner–Brazler shear test. Journal of the Science of Food and Agriculture. 79(12): 1737–1746.
Velazquez, G., Miranda, P., López, G., Vázquez, M., Torres, J.A. and Ramirez, J.A. 2007. Effect of recovered soluble proteins from pacific withing surimi wash water on functional and mechanical properties of Alaska Pollock surimi grade FA. Ciencia y Tecnología Alimentaria. 5(5): 340–345.
Vestergaard, M., Madsen, N.T., Bligaard., H.D., Bredahl, L., Rasmussen, P.T. and Andersen, H.R. 2007. Consequences of two or four months of finishing feeding of culled dry dairy cows on carcass characteristics and technological and sensory meat quality. Meat Science. 76: 635–643.
Voisey, P.W. 1976. Engineering assessment and critique of instruments used for meat tenderness evaluation. Journal of texture studies. 7: 11–48.
Voisey, P.W. and Nonnecke, I.L. 1971. Measurement of pea tenderness. Journal of Texture Studies. 2(3): 348–364.
Volodkevich, N.N. 1938. Apparatus for measuring of chewing resistance or tenderness of foodstuffs. Journal of Food Science. 3: 221–225.
Warner, K.F. 1928. Progress report of the mechanical test for tenderness of meat. Proceeding of the American Society Animal Production. 21: 114–118.
Wheeler, T.L., Koohmaraie, M. and Shackelford, S.D. 1995. Standardized Warner-Bratzler shear force procedures for meat tenderness measurement. Proc. Recip. Meat Conf. 50: 68–77.
Winkler, C.A. 1939. Tenderness of meat: I. A recording apparatus for determining the tenderness of meat. Canadian Journal Reaserch. 17: 8–14.
Xiong, R., Cavitt, L.C., Meullenet, J.F. and Owens, C.M. 2006. Comparison of Allo-Kramer, Warner-Bratzler and razor blade shears for predicting sensory tenderness of broiler breast meat. J. Texture Studies 37: 179–199.
Zhang, M., Wang, D., Huang, W., Liu, F., Zhu, Y., Xu, W. and Cao, J. 2013. Apoptosis during postmortem conditioning and its relationship to duck meat quality. Food Chemistry. 138: 96–100.
Zhuang, H. and Savage, E.M. 2009. Variation and Pearson correlation coefficients of Warner-Bratzler shear force measurements within broiler breast fillets. Poultry Science. 88: 214–220.

5

Pigments in Fruit and Vegetables

Sara M. Oliveira, Cristina L.M. Silva and Teresa R.S. Brandão*

ABSTRACT

Fruit and vegetables are important sources of nutrients with beneficial effects on human health such as the prevention or delaying of diseases. Many of these effects are promoted by the antioxidant properties of pigments, which are responsible for the color of the products. The diversity of colors and biological properties of fruit and vegetables are related with the chemical structures of the synthetized pigments that are generally classified as carotenoids, tetrapyrroles, polyphenolic compounds and alkaloids.

The identification and quantification of natural pigment compounds are important for identifying foods with functional properties, which may be related to health-promotion or disease prevention. However, research on this topic involves many analytical methods and requires a deep knowledge in diverse scientific areas.

This chapter provides an overview of the analytical procedures developed for pigment analysis in fruit and vegetables, including a brief summary of the relevant extraction techniques often used. Some technologies, which in the past were more restricted to basic research, are also highlighted.

CBQF—Centre of Biotechnology and Fine Chemistry - State Associated Laboratory, Faculty of Biotechnology, Catholic University of Portugal, Porto, Rua Dr. António Bernardino Almeida, 4200-072 Porto, Portugal.
* Corresponding author

5.1 Introduction

Society today is increasingly concerned with nutrition and as such a healthy diet plays a fundamental role in the achievement of physical and mental well-being. The concept of functional food components emerged and, within this framework, fruit and vegetables have risen as significant contributors, due to their intrinsic beneficial effects, providing not only the basic nutritional requirements, but also acting on the prevention or delaying of diseases. Many of these effects are promoted by the anti-oxidant properties provided by pigments, contained inside the cells and responsible for the color. Color is one of the most important characteristics of a product, since it is directly related to its appearance, constituting one of the primordial quality attributes assessed by the consumer. The chemical structure of the pigments synthetized by fruit and vegetables is the basis for the diversity of colors and biological properties of these products. According to their chemical structures, plant pigments are categorized into four groups: carotenoids, tetrapyrroles, polyphenolic compounds, and alkaloids (Fig. 5.1) (Schoefs 2004).

Carotenoids are localized in chloroplasts and chromoplasts, playing a crucial role in photosynthesis (Khoo et al. 2011). Carotenoid pigments consist of two major classes: the hydrocarbon carotenes and the oxygenated xanthophylls (Schwartz 1998). Although more than 600 carotenoids have been identified, it is estimated that only about 40 carotenoids can be absorbed, metabolised

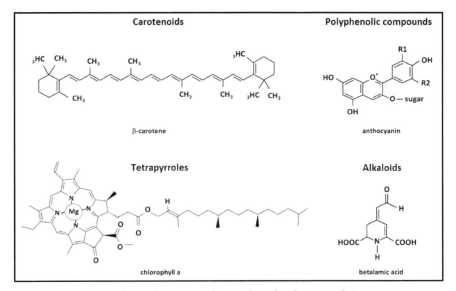

Figure 5.1 Chemical structure of a member of each group of pigments.

and/or used by our bodies. This number decreases to six, considering the carotenoids usually detected in human blood plasma. This group includes α- and β-carotene, lycopene, β-cryptoxanthin, zeaxanthin and lutein. Their major sources are lettuce, spinach, Brussels sprouts, pepper, pumpkin, tomato, carrot, grape, mango, melon, orange and watermelon, among others (Fernández-García et al. 2012). Several reports have demonstrated that carotenoids provide yellow to red color in many fruit and vegetables (Bartley and Scolnik 1995; Lancaster et al. 1997; Hornero-Méndez and Mínguez-Mosquera 2000), depending on conjugated double bonds and on the functional groups of the molecule (Khoo et al. 2011).

One of the most important biological activities of some carotenoids is the presence of a β-type non-substituted ring that enables animals and humans to metabolise them into retinol, i.e., vitamin A (Fernández-García et al. 2012). In fact, these retinol precursors can only be obtained from food sources of which fruit and vegetables have the most prominent pro-vitamin A activity.

Tetrapyrroles include chlorophylls *a* and *b*, photosynthetic pigments that contribute to the characteristic green color of diverse vegetables, such as broccoli and cabbage plants. Post-harvest senescence of green vegetables may be assessed by the chlorophylls composition of the cell lipid membranes. The oxidative activity resultant of chlorophyll deterioration leads to membrane deterioration, loss of compartment and cell breakdown (Deschene et al. 1991; Zhuang et al. 1995). The chlorophylls include the basic skeleton structure of porphyrin with a magnesium ion in the center and a long phytol group in the tail (Schwartz and Lorenzo 1990). These pigments are very effective photoreceptors because they contain a network of alternating single and double bonds; the electrons become delocalized and stabilise the structure (Fernández-León et al. 2010). Such delocalised polyenes have very strong absorption bands in the visible regions of the spectrum, allowing the plant to absorb the energy of the sunlight (Streitweiser and Heathcock 1981). However, plants are unable to assimilate all of the stored energy, being considered low-efficient organisms (Fernandez-Jaramillo et al. 2012). The excess of energy absorbed by the leaves is dissipated through different mechanisms involving thermal dissipative process, fluorescence emissions and photochemistry (Losciale et al. 2011). These processes have been employed as useful tools for the development of non-destructive methodologies for the assessment of quality and the chemical or physical characteristics of fruit and vegetables. Most of these methods rely on the chlorophyll fluorescence, defined as the red and far-red light emitted by photosynthetic tissue when it is excited by a light source, and have been used to explain certain physiological behaviors in plants, such as photochemical and non-photochemical quenching (Fernandez-Jaramillo et al. 2012).

Polyphenols are a structural class of organic chemicals characterized by the presence of large multiples of phenol units. Anthocyanins belong to a group of flavonoids, a subclass of polyphenols responsible for the red, blue, and purple colors of some fruit, vegetables, juices and wines. These compounds are natural water-soluble pigments constituted by a basic skeleton, often acylated with one or more polar side chains such as glucosides (Schoefs 2004). In addition, and depending on pH, the oxygen atom of the hetero-cycle may be positively charged, resulting in structural transformations, a unique aspect of most anthocyanins (Schwartz 1998). The differences between individual anthocyanins are related to the number of hydroxyl groups, the nature and number of sugars attached to the molecule, the position of this attachment, and the nature and number of aliphatic or aromatic acids attached to sugars in the molecule (Kong et al. 2003). Over 500 different anthocyanins have been isolated from plants (Yang et al. 2011). Red wine, vegetables and fruit represent the main contributors of anthocyanins to the human diet. Among the vegetables, cabbages, beans, onions and radishes are the most representative; in the fruit category, berries, strawberries, oranges, acerola and açaí were described as rich anthocyanin sources (Dugo et al. 2001; de Rosso 2008; Yang et al. 2011). Increasing attention has been given to these pigments due to their health benefits provided by the antioxidant and anti-inflammatory properties that are conferred by the scavenging capacity against superoxide radicals, hydrogen peroxide, hydroxyl radicals, and singlet oxygen (Wang et al. 1999; Wang and Jiao 2000). Additional therapeutic attributes were reported, once anthocyanins act as radiation-protective, chemoprotective and vasoprotective agents, decreasing the risks of cardiovascular diseases (Wang et al. 1997; Seeram and Nair 2002). Their protective efficiency depends on the chemical structure of the molecule, such as degree of glycosylation and number of hydroxyl groups in the B ring (Kong et al. 2003). Furthermore, food industry has been using anthocyanins as natural food colorants to replace the original color that is lost during processing (Longo and Vasapollo 2005).

Alkaloids are a group of naturally occurring organic nitrogen-containing bases. Alkaloids are found primarily in plants and are especially common in certain families of flowering plants. Belonging to the alkaloids group, betalains are natural water-soluble pigments, which are synthesized from the amino acid tyrosine, and may be divided into two structural groups: the red-violet betacyanins and the yellow-orange betaxanthins (Azeredo 2009). Betalamic acid is the chromophore common to all betalain pigments, and the pigment classification as betacyanin or betaxanthin is determined by the nature of the betalamic acid addition residue (Strack et al. 2003). These pigments are not widely distributed throughout the plant kingdom and their synthesis is restricted to the taxonomic group of *Centrospermeae* species (Schoefs 2004). Purple red beet root contains high concentrations of betalain pigments, which consist of the predominant purple-red betacyanins and lower concentration of the yellow betaxanthins (Schwartz 1998). Betalains are synthesized in the cytoplasm and then stocked in vacuoles, being found mainly in flowers and fruit and occasionally in vegetative tissues of plants (Pavokovic and Krsnik-Rasol 2011). Red beet is the major betalain source for commercial purposes, which produces mainly betanin, resulting in poor color variability (Stintzing and Carle 2004). However, a wider color palette may be produced when combined with intensely colored juices (Girod and Zryd 1987). In Europe, this pigment is listed as E162 and has been employed in several processed foods such as dairy products, frozen desserts and meat (Stintzing and Carle 2007). Currently, the main sources of these pigments are crop plants, but efforts have been made to obtain similar yields from alternative sources, such as cell suspensions or hairy roots grown in bioreactors (Clement and Mabry 1996). Given their economic and health benefits there is an increasing level of interest into betalains that is also promoted by their profitable properties from the chemical, medical and pharmacological points of view. Besides being chemically stable over a wider pH range than anthocyanins, studies have shown their inhibitory efficiency on cancer (Kapadia et al. 1996; Kapadia et al. 2003), excellent radical scavenging activity (Matysik et al. 2002), powerful antioxidant activity and great anti-inflammatory response (Stintzing and Carle 2004). Moreover, red beet may be related with increased pathogen resistance (Mabry 1980; Piattelli 1981; Stafford 1994) and improved viral defense (Sosnova 1970).

For all of the above-mentioned reasons, it is increasingly necessary to develop methodologies for pigment analysis in natural products, such as fruit and vegetables. Due to the complexity of such samples, analysis is not straightforward and requires intermediate steps in order to extract, separate, purify and/or characterize, prior to analysis and quantification. Furthermore, fresh foods require controlled processing and storage conditions to minimize modifications concerning pigments content in the final product. These changes, as well as chemical modifications, may affect the color and taste of the product, affecting the quality and causing a negative impact on the consumer. To overcome this, it is extremely important that food scientists join forces seeking the development of simple, fast and accurate methodologies for pigments quantification.

This chapter provides an overview of the analytical procedures developed for pigment analysis in fruit and vegetables, including a brief summary of the extraction techniques.

5.2 Pigments Extraction

Extraction of pigments from biological samples is required prior to identification and quantification. This may be a challenging task, since it is often time-consuming and requires several steps that may affect the precision of the analysis. In addition, it may be tedious and harmful to the analyst, as in some cases complete extraction is only attained through the use of several solvents, especially when pigments with different polarities are extracted from the same sample.

5.2.1 Carotenoids and Chlorophylls

Chlorophyll and carotenoid molecules are usually hydrophobic compounds and, therefore, they can be extracted with a single or a mixture of organic solvents using a homogenizer (Schoefs 2004).

Taungbodhitham et al. (1998) evaluated three extraction methods for the analysis of carotenoids in fruit and vegetables, showing that ethanol and hexane, two solvents of low biological hazard, were the most suitable for extracting carotenoids from the matrix. The use of double extraction with a mixture of ethanol and hexane also provided good recoveries of lycopene, α-carotene and β-carotene, when added to tomato juice, carrot and spinach.

Saponification is also referred to as a method for the extraction of carotenoids from fruit and vegetables (Konings and Roomans 1996), using portions of methanol/tetrahydrofuran (1:1 v/v) at 0 °C until colorless. After addition of NaCl 10% (w/v), unsaponified and saponified solutions were extracted with petroleum ether, until petroleum ether phase was colorless. From the saponified samples, combined petroleum ether portions were washed with water until reaction was neutral. Organic layers were evaporated to dryness and the residue was dissolved by ultrasonic agitation in methanol/tetrahydrofuran (75:25 v/v). When necessary, the extract was saponified at room temperature for two hours with an equal volume of KOH in methanol 10% (w/v). Although saponification has been recommended only if strictly necessary, some authors modified the standard procedure to include this process. In mango samples, the separation of all-*trans*-β-carotene and its *cis* isomers was improved when compared to the unsaponified extract (Vásquez-Caicedo et al. 2005). Saponification was also used to eliminate interferences of other pigments such as chlorophylls (Müller 1997; Dachtler et al. 2001; Aman et al. 2005a).

The use of extraction/purification cartridges packed with sorbents was also described for carotenoid extraction. Gentili and Caretti (2011) extracted carotenoids from kiwi fruit, using C18 sorbent, in which organic extracts were forced to pass through it by a water-pump vacuum. Lutein and zeaxanthin were extracted from spinach samples using C30 sorbent material and a few crystals of butylated hydroxytoluene used as stabilizer (Glaser et al. 2003).

Regarding chlorophylls extraction, the use of organic solvents is the most popular procedure. In some cases, repeated extraction steps were performed prior to identification analysis (Edelenbos et al. 2001; Kao et al. 2011). Bohn and Walczyk (2004) submitted extracts of spinach, lettuce and endive to an additional purification step, based on solid-phase extraction.

5.2.2 Anthocyanins

Established anthocyanin extraction methods involve maceration (Revilla et al. 1999) or soaking the plant material in a low boiling point alcohol (e.g., methanol) containing a small amount of mineral acid, thus maintaining a low pH to keep the anthocyanins in their stable flavylium form (Garcia-Viguera et al. 1998). The use of methanol acidified with HCl was selected for extraction of anthocyanins from beans (Takeoka et al. 1997; Macz-Pop et al. 2006), berries (Nyman and Kumpulainen 2001; Longo and Vasapollo 2005), grapes (Mazzuca et al. 2005), camu-camu (Zanatta et al. 2005), acerola and açaí (de Rosso et al. 2008; Bordonaba et al. 2011). Acetic acid was the extractant applied in sweet potato cell suspension cultures (Konczak-Islam et al. 2003; Plata et al. 2003).

Nevertheless, Anderson et al. (1970), who identified acetic acid as an acylating agent of anthocyanin pigments in grapes, reported that facile hydrolysis of anthocyanin acetates occurs upon the exposure to trace quantities of mineral acid during the extraction. Some authors, working on chemotaxonomic studies, have reported that the extraction of some acylated anthocyanins under acid conditions may cause their partial or total hydrolysis (Revilla et al. 1998). Therefore, in order to obtain anthocyanins closer to their natural state, several solvents have been suggested for the initial pigment extraction, such as acetone, formic acid in methanol or water, mixtures of methanol-acetic acid-water and hydrochloric acid in methanol (Garcia-Viguera et al. 1998).

Revilla et al. (1998) developed a rapid method, avoiding the use of acid extraction solvents, for the extraction of anthocyanins from grapes and compared several extraction procedures with different solvents (acid vs. neutral). Each procedure involved two or more successive extractions, and all of them included, as a first step, the immersion of the sample in pure methanol, containing an adequate proportion of 12 N hydrochloric acid, if necessary, and followed by grinding in a blender for one minute. When each extraction step was finished, the liquid extract was separated by centrifugation, and the residue was submitted to extraction again. This work pointed out that the use of solvents containing up to 1% of 12 N hydrochloric acid causes partial hydrolysis of some acetylated anthocyanins during extraction. Consequently, the use of these solvents should be avoided in grapes or in any other plant material that contains acetylated anthocyanins.

Garcia-Viguera et al. (1998) evaluated the use of acetone as an extraction solvent for anthocyanins in strawberries, and compared it to other widely used solvents, such as formic acid in methanol or water, a mixture of methanol/acetic acid/water, and hydrochloric acid in methanol. Both formic acid in methanol and acetone originated pectin clotting and a complete separation of liquor from the solid mass of the fruit, producing a very clear liquid after filtration. The other solvents formed a turbid extract that had to be passed through filters in several steps. Additionally, the use of acetone provided higher accuracy values in the quantitative determinations than those obtained using other solvents.

Recently, different solvents were also evaluated on jabuticaba anthocyanin pigments (Lima et al. 2011). The first extraction of the skin pigments was performed with HCl 0.01 mol L^{-1} solution. Maceration for 12 h at 4 °C in acidified water with HCl 1.5 mol L^{-1} and 0.01 mol L^{-1} was also tested, as well as with other solvents, such as 95% and 50% ethanol. The highest extractability of anthocyanins was attained with the latter, followed by 30% acetone and 50% methanol. Extraction with water resulted in the lowest extractability of anthocyanins. This allowed concluding that in spite of anthocyanins being water-soluble, the less polar solvents removed higher amounts of pigment from the sample. The authors described as the best method for extracting jabuticaba anthocyanins, maceration using 50% ethanol and/or 95% ethanol acidified with HCl 1.5 mol L^{-1} (85:15) using the proportion of 1:15 (1g sample:15mL solvent). Using this approach, a low pH extract was obtained, and the anthocyanins were stable over a period of 185 days.

Arapitsas et al. (2008) resorted to a Dionex accelerated solvent extractor, for anthocianins extraction from red cabbage leaves, and used a mixture of water/ethanol/formic acid (94:5:1 v/v/v) as the extraction solvent. They separated and identified twenty-four anthocyanins, all having cyanidin as a glycon, represented as mono- and/or di-glycoside, and acylated, or not, with aromatic and aliphatic acids. In addition, and for the first time, the procedure allowed the identification of nine anthocyanins in red cabbage.

Separation and purification of anthocyanins from berries (Dugo et al. 2001), strawberries (Andersen et al. 2004), raspberries and grapes (Tian et al. 2005), violet cauliflower and red cabbage (Scalzo et al. 2008) and other fruit and vegetables (Can et al. 2012) by chromatographic techniques, as well as the use of solid phase extraction with C18 resin (Giusti et al. 1999; Garzón et al. 2009) were also reported.

5.2.3 Betalains

Because betalains and anthocyanins are chemically (but not biochemically) related, extraction and pre-purification methods used for anthocyanins can also be applied to the preparation of betalains (Schoefs 2004).

Products containing these pigments are often macerated or ground. Betalain pigments are ionic, exhibit high water solubility, and are therefore extracted from plant tissues with water (Schwartz 1998). Castellar et al. (2006) reported better extraction of betalains from Opuntia fruit using water in comparison with a mixture of ethanol and water. However, in most cases, the use of methanol or ethanol solutions (20–50%) is required for complete extraction (Delgado-Vargas et al. 2000). Higher pigment levels were found with ethanol/HCl (99:1 v/v) than with water, although better betalain stability was achieved with aqueous extraction (Azeredo 2009). Pigment stability enhancement and avoidance of oxidation by polyphenoloxidases can be attained by slight acidification of the extraction medium (Schliemann et al. 1999; Strack et al. 2003). Enzimatic treatments were also described to avoid betalain enzymatic degradation, favoring betalain extraction (Delgado-Vargas et al. 2000; Moβhammer et al. 2005).

The degree of cell membrane permeabilisation affects the extraction efficiency (Azeredo 2009). Pulsed electric field and gamma irradiation pretreatments were reported to increase cell permeability, providing higher extraction efficiencies (Rastogi et al. 1999; Ade-Omowaye et al. 2001; Chalermchat et al. 2004; Fincan et al. 2004; Nayak et al. 2006). On the other hand, irradiation also increased betanin degradation rates (Azeredo 2009).

Several processes, including diffusion-extraction (Wiley and Lee 1978), solid–liquid extraction (Lee and Wiley 1981), reverse osmosis (Lee et al. 1982) and ultrafiltration (Bayindirli et al. 1988) were described to be efficient in recovering betalains from beet tissue. However, recently developed methodologies exploit chromatographic tools for purification of samples and pigments separation, after removal of pectins by precipitation with ethanol (Herbach et al. 2006; Rebecca et al. 2010) and/or phenolic compounds (Stinzing et al. 2004).

5.3 Methodologies for Pigments Assessment

Suitable procedures are necessary to separate and identify pigments from fruit and vegetables, after extraction. The selection of the appropriate strategy is strongly conditioned by several factors, such as nature and amount of the sample material available, properties of the analytes, nature of the interfering species present in the sample matrix, type of results (qualitative or quantitative) required, admissible duration and cost of the analysis. In this context, chromatographic separation techniques have been playing an important role, and improvements have been made to accompany the nutritional findings, constantly renewed by food scientists. However, chromatographic-based techniques have some limitations and non-chromatographic methodologies have been reported as alternatives.

5.3.1 Chromatographic Methods

High Performance Liquid Chromatography

High resolution, speed, reproducibility and sensitivity are some of the main advantages offered by the high-performance liquid chromatography (HPLC) systems, constituting therefore the most commonly used methodology to quantify pigments from fruit and vegetables. With this purpose, several HPLC methods using diverse eluents, column phases and detection systems have been described, and some of them will be presented in this section.

Although normal-phase HPLC has been described for separation and identification of pigments (Eder 2000), reverse-phase HPLC is the most widely used technique for routine analysis, justified by the large availability of reverse phases. Simple solvent systems based on methanol, water and ethyl acetate may be used in conjunction with modern stationary phases, in which retention is largely affected by the analyte polarity (Scotter 2011). Photosynthetic pigments, chlorophylls and carotenoids have a clear hydrophobic character and are usually analyzed on C18-reversed-phase (RP) columns (Schoefs 2004). The C30-RP columns are particularly efficient for the separation of carotenoids because the interactions of the pigments and the stationary phase are maximized by their similar size (Scotter 2011). This characteristic enables the separation of carotenoid isomers in vegetables by using a C30-RP column (Emenhiser et al. 1995; Lacker et al. 1998; Lee and Chen 2001; Aman et al. 2005b). A C34 alkyl-bonded stationary phase was also evaluated, but only a slight improvement was achieved when compared with C30 stationary phases. However, and for some isomers, a better resolution was obtained due to the ability of the large carotenoid molecules to fully interact with the C34 phase (Bell et al. 1996). On the other hand, the use of calcium hydroxide columns provided good resolution in the separation of carotene geometric isomers (Schmitz et al. 1995).

The mobile phases selected for the separation of hydrophobic molecules by HPLC systems are usually based on a mixture of polar and non-polar solvents with or without water, depending upon the separation required. The use of a polar organic solvent, mixed with a small amount of water is recommended when polar molecules, such as glycosyl esters of carotenoids or chlorophyll *c*, are present in the mixture (Schoefs 2004). Moreover, this strategy is also applied for the separation of the more hydrophilic compounds, such as anthocyanins and betalains. Isocratic elution is usually adequate for separation of a small number of target analytes, but gradient elution with two or more solvent systems is often required for the separation of large numbers of carotenoids covering a wide range of polarities (Scotter 2011). Separation and quantification of anthocyanins is achieved by using mostly acetonitrile as the mobile phase. A new acetonitrile-free mobile phase for determination of individual anthocyanins in blackcurrant and strawberry fruit was presented by Bordonaba et al. (2011). The authors studied the potential of a methanol-based mobile phase that could readily replace the widely used acetonitrile. They concluded that the proposed approach allowed appropriate elution and quantification with comparable accuracy to the one that used acetonitrile as a mobile phase. The advantages are related with the recent shortage of acetonitrile, and its current prices.

Pól et al. (2004) developed a method for determination of lycopene in food by on-line solid phase extraction (SPE) coupled to HPLC using a single monolithic column for trapping and separation. Lycopene was not exposed to atmospheric air or light during the analytical procedure, and unwanted and unexpected degradation was thereby eliminated.

Pigment separation may be enhanced by heating the column (Choung et al. 2003), fractionating the different groups of pigments by using an open column packed with alumina (Mercadante et al. 1998) or purifying the extract on a C18-RP cartridge previously activated with acidified methanol (Hong and Wrolstad 1990; Palozza and Krinsky 1991; Takeoka et al. 1997), prior to HPLC analysis of samples.

HPLC methodologies used for the determination of pigments in foods and vegetables have been coupled to diverse detection systems, such as UV-Visible spectroscopy, infrared spectroscopy, fluorimetry, mass spectroscopy, nuclear magnetic resonance, Raman spectroscopy and photoacoustic spectroscopy. An overview of these methodologies was recently presented by Schoefs (2005). However, since then, several new methods were proposed and some examples will be described below.

Regarding UV-Visible spectroscopy, diode array detectors (DAD) allow *in situ* identifications immediately after separation. HPLC methods with UV/Vis detectors are very popular for determination of carotenoids, and applications to raw vegetables and fruit were reported: tomatoes, watermelon and grapes (Cucu et al. 2012), tomato (Kotíková et al. 2009; Dias et al. 2008; Kotíková et al. 2011), Indian vegetables (Aruna et al. 2009), orange, pear, peach, apple, cherry, Portuguese coles, turnip greens, purslane, leaf beet and beetroot leaves (Dias et al. 2009), strawberry, raspberry, blackberry, blueberry, black currant and red currant (Marinova and Ribarova 2007), papaya and peach (Sentanin and Amaya 2007), Asian vegetables (Kidmose et al. 2006), carrot, pepper, watermelon, persimmon and medlar (Barba et al. 2006) and wolfberry (Peng et al. 2005). HPLC with photodiode array detector (PDA) has proven to be a useful tool for the characterization of anthocyanins in blackcurrant, strawberry fruit (Bordonaba et al. 2011), cranberries (Brown et al. 2011), raspberry (Bononi et al. 2006) and beans (Macz-Pop et al. 2006). However, this procedure requires authentic standards, which are often difficult to obtain and therefore quantification of anthocyanins has been carried out using only one reference compound (Zanatta et al. 2005).

Can et al. (2012) developed a rapid and feasible assay procedure for the determination of free forms of the six most abundant anthocyanins in foods, including fruit and vegetables. The pigments were separated by gradient elution and quantified using HPLC-DAD. A fast sample preparation step was developed, which allowed direct injection of samples to the chromatograph without the need of chemical extraction. The use of photodiode array and fluorimetric detectors also allows characterization and identification of chlorophyll, followed by HPLC separation (Scotter 2011). Mendiola et al. (2010) developed a method for profiling the compounds responsible for the antioxidant activity in vegetable extracts, allowing the simultaneous determination of water-soluble vitamins, fat-soluble vitamins, phenolic compounds, carotenoids, and chlorophylls in a single run by HPLC-DAD.

The application of mass spectrometry to HPLC pigment analysis is also a common strategy. The main advantage of HPLC with mass spectrometry detection is that it enables not only the analyte quantification, but also the elucidation of its structure based on the molecular mass and fragmentation (Scotter 2011). The application of mass spectrometry to pigment analysis was reported for the determination of carotenoids in tomatoes (Lucini et al. 2012) and melons (Fleshman et al. 2011), and also for detection of anthocyanins in cauliflower (Scalzo et al. 2008).

Diode array detectors may be combined with mass spectrometry. This approach was used for the determination of chlorophylls in anti-carcinogenic Chinese herbs (Loh et al. 2012), anthocyanins in *Berberis boliviana* Lechler fruit (Jiménez et al. 2011) and acerola (de Rosso 2008). The use of a mass detector was crucial to recognize 12 different compounds in English cherries (Ieri et al. 2012). The experimental conditions used allowed determination of phenolic acids, anthocyanins and coumarins.

Electrospray ionization (ESI) and atmospheric-pressure chemical ionization (APcI) are ionization modes commonly used in mass spectrometry. The sensitivity of the ESI mode is generally two orders of magnitude higher than that of UV-Vis detection, thus a detection limit close to 5×10^{-10}g can be achieved for β-carotene (Feltl et al. 2005). This mode was applied in the determination of anthocyanins in blood orange (Cao et al. 2009), Andes berry fruit (Garzón et al. 2009), red cabbage (Arapitsas et al. 2008) and sweet cherry (Sugawara and Igarashi 2008).

Most of the methods applied to carotenoids use APcI mode due to the absence of protonation sites in these molecules (Feltl et al. 2005). These procedures were applied for the detection of carotenoids in peach (Giuffrida et al. 2013), fruit of Araza (Garzón et al. 2012) and mango fruit (Ornelas-Paz et al. 2008). Anthocyanin determination in orange (Cao et al. 2009) was also reported.

Herbach et al. (2006) presented a method based on isotope ratio determination for betalain purification from red beets and purple pitaya. Betanin and isobetanin were isolated by combining different column chromatography techniques with semi-preparative HPLC. According to the authors, determination of the carbon isotope ratios of betanin and isobetanin was found a feasible approach to authenticity evaluation of products derived from purple pitaya and red beet.

HPLC has also been used to evaluate the purity of commercially available food colorants derived from natural foods, such as chlorophyllin (Chernomorsky et al. 1997; Schoefs 2002).

Open Column Chromatography

Open column chromatography (OCC) was described for chlorophyll separation, employing phases such as powdered sucrose, DEAE-Sepharose, cellulose or MgO/Hyflosupercel (Wesley et al. 1970; Strain et al. 1971; Omata and Murata 1983). Lutein from Indian vegetables and vegetable oils was purified by open column chromatography on neutral alumina, using hexane extract (Aruna et al. 2009). The β-carotene was eluted with hexane, the lutein and zeaxanthin fraction was eluted with methanol/dichloromethane (1:1 v/v), and the fractions rich in violaxanthin and neoxanthin were eluted with ethyl acetate/hexane (5:5 v/v) and ethyl acetate/hexane (1:9 v/v), respectively. The purity of lutein and zeaxanthin elute was analyzed by HPLC and their respective spectra and absorption maxima were confirmed by photodiode array (PDA) detector. Anthocyanins were also separated by OCC, but these methods were applied to wine analysis (Schoefs 2004).

Thin Layer Chromatography

Thin layer chromatography (TLC) is often used for the separation and analysis of pigments in foods and, due to the short analysis time, low cost and simplicity; it is frequently used for qualitative purposes as a screening tool. This technique allowed the identification of 13 carotenoids from yellow passion fruit (Mercadante et al. 1998), which identity was confirmed by mass spectrometry (MS) and nuclear magnetic resonance (NMR) analysis. Qualitative pigment differences between fresh (unripe and ripe), frozen, and canned kiwi fruit slices were assessed by TLC, HPLC, UV-Visible spectroscopy and chemical tests (Cano and Marín 1992). The principal pigments in fresh and frozen kiwi were xantophylls, chlorophylls and their derivatives. The pigment pattern of the canned product showed the formation of zinc complexes and thermal treatment induced the degradation of some xantophylls, violaxanthin and neoxanthin. Hornero-Méndez and Mínguez-Mosquera (1998) employed silica gel plates in a pre-saturated chamber to separate carotenoids from *Capsicum annuum*. The solvent system was the mixture petroleum ether/acetone/diethylamine (10:4:1 v/v/v). However, when the separation is achieved on silica gel, it is necessary to first neutralize the acidity of silica before pigment separation, because its slight acidity may trigger pigment degradations such as epoxide–furanoxide rearrangement in carotenoids and chlorophyll-pheophytinization (Schoefs 2004). Although it is mostly applied to carotenoids, separation of Cu(II) complexes of oxidized chlorophylls and allomerization products of pheophytins *a* and *b* from fresh spinach leaves was also reported (Mínguez-Mosquera et al. 1996). Betalain pigments from red beet extracts were also separated by TLC (Bilyk 1981). Henning (1983) used reversed phase solid-phase extraction (SPE) to isolate betanin from food extracts, which was then subjected to TLC on cellulose, with butanol/formic acid/water (10:3:3) as developing solvent. Separation by TLC has been used mainly to separate fractions containing groups of compounds of similar polarity. The solid phases often used are sucrose, DEAE-sepharose or cellulose (Scotter 2011). Petroleum ether with acetone or hexane were the principal mobile phases for separation of carotenoids from vegetable products (Schwartz and Patroni-Killam 1985; Kanasawud and Crouzet 1990; Meléndez-Martínez 2006; Ren and Zhang 2008). TLC features may be improved by using high performance thin layer chromatographic (HPTLC) plates. Daurade-LeVagueresse and Bounias (1991) separated chlorophylls, pheophytins *a* and *b*, β-carotene, lutein, violaxanthin and neoxanthin from barley leaves on HPTLC CN-coated (i.e., carbon-nitrogen coated) sheets using chloroform/hexane/methanol (25:70:5 v/v/v). Suzuki et al. (1987) developed a reverse phase HPTLC methodology for determination of chlorophylls *a* and *b*, pheophytins *a* and *b* and pheophorbides *a* and *b*, using an octadecyl-bonded stationary phase with mobile phase solvents of different polarities, including alcohols, acetone, acetonitrile and mixtures of ethanol and water.

The selection of phases and poor resolution on the separation of compounds with similar structures represent the main limitations of TLC methods. However, they are a useful tool for the separation of carotenoids from crude extracts and allow obtaining relatively large amounts of pigments, prior to analysis by HPLC (Isaksen and Francis 1986; Schiedt 1995; He et al. 1998; Delgado-Vargas et al. 2000).

Counter Current Chromatography

Counter current chromatography (CCC) is a procedure alternative to solvent extraction. It allows the analysis of large-scale sample preparations, since it is an automated version of liquid–liquid extraction, comparable to the repeated partitioning of an analyte between two immiscible solvents by vigorous mixing in a separatory funnel (Schoefs 2004). Countercurrent chromatography operates under gentle conditions and allows a nondestructive isolation of labile natural compounds. Due to the absence of any solid stationary phase, adsorption losses are minimized and, consequently, the sample recovery is close to 100% (Schwarz et al. 2003).

Despite the intrinsic advantages of CCC, its use as a routine technique for pigment analysis in fruit and vegetables is still limited, with only two works available in literature. One describes the use of CCC for the isolation of anthocyanins from purple corn, elderberry juice, and blackberries, among other foodstuffs (Schwartz et al. 2003). High amounts of pure anthocyanins were obtained within a single CCC run, using sample loads between 500 and 1000 mg. The chromatograph was equipped with three coils connected in series and three solvent systems were used. Solvent system I consisted of n-butanol/methyl tert-butyl ether/acetonitrile/water (2:2:1:5 v/v/v/v, acidified with 0.1% trifluoroacetic acid). Solvent system II had the same composition but contained less trifluoroacetic acid (0.01%). Solvent system III was n-butanol/methyl tert-butyl ether/acetonitrile/water (3:1:1:5 v/v/v/v, acidified with 0.1% trifluoroacetic acid). The less dense layer was always used as the stationary phase and therefore the elution mode was head-to-tail. Isolated pigments included monoglycosylated, acylated and highly glycosylated derivatives of anthocyanins. The other work reports the use of CCC to isolate chlorophylls *a* and *b* from spinach (Jubert and Bailey 2007). The solvent system comprised two phases (heptane/ethanol/acetonitrile/water, 10:8:1:1 v/v/v/v), which were mutually saturated by shaking in a separatory funnel and separated immediately before use. The aqueous bottom layer served as the CCC mobile phase to obtain highly pure chlorophyll *a*, while the top layer (rich in heptane) served as the CCC mobile phase to obtain highly pure chlorophyll *b*. Both methods required purity and identity confirmation of the isolated pigments by HPLC, mass spectrometry and nuclear magnetic resonance.

Supercritical Fluid Chromatography

Supercritical fluid extraction (SFE) is a rapid and selective method that lends itself to automation (Scotter 2011). Its main advantages are: rapidity, control of the solvent strength, and the solvents used are environmentally friendly gases with low toxicity (Delgado-Vargas et al. 2000). With this method the concentration stage is avoided, because solvents are immediately eliminated under environmental conditions (Chester et al. 1992; Hawthorne 1990). Because of its critical temperature and its nontoxic, nonflammable, and environmentally preferred qualities, supercritical carbon dioxide (SC-CO$_2$) is widely used as solvent in the supercritical fluid extraction (Gnayfeed et al. 2001). Carotenoids were extracted from carrots using SC-CO$_2$ fluid extraction (Barth et al. 1995; Vega et al. 1996). Ethanol concentration and temperature are the most relevant factors affecting the extraction. Total vitamin A activity (α- plus β-carotene) was higher in SFE than in traditional solvent extraction and the extraction time was reduced from six to one hour (Barth et al. 1995). Supercritical CO$_2$ to extract β-carotene of sweet potato provided approximately a five-fold or a three-fold increase in the amount of carotenoids extracted from freeze-dried tissue relative to the amount extracted from oven-dried or fresh tissue, respectively (Spanos et al. 1993). This methodology was also applied in the determination of carotenoids in hot spice red pepper, known as paprika (Jarén-Galán et al. 1999; Gnayfeed et al. 2001; Ambrogi et al. 2002). Gnayfeed et al. (2001) evaluated the use of subcritical propane at different conditions of pressure and temperature to estimate the yield and variation in carotenoid, tocopherol, and capsaicinoid contents and composition. They concluded that although propane in subcritical conditions was an excellent solvent for the recovery of carotenoid pigments and tocopherols, it barely solvated capsaicinoids from paprika. This suggests the use of mixtures of CO$_2$ and propane to produce paprika oleoresins with different color intensities, pungency levels, and antioxidant contents. Ambrogi et al. (2002) studied the fractionation of pigments that occurs during supercritical extraction of oil from natural matrices containing carotenoids. They reported a total carotenoid recovery of 75%, using only 15% of the total CO$_2$ required for the total extraction of the pigments, under controlled pressure (30MPa), temperature (60 °C) and CO$_2$ flow (8–9 kg/hr). No noticeable difference was observed in carotenoid concentration when paprika powder and paprika oleoresin were compared, suggesting that the fractionation process was independent of the cellular structure.

Capillary Electrophoresis

Capillary electrophoresis allows fast separations of the components at relatively low cost, because the amount of solvent and waste are strongly reduced. The good separation and fast elution of the pigments suggest that capillary electrophoresis is suitable for routine analysis of tiny samples (Schoefs 2004). Electrophoretic techniques, including paper electrophoresis and capillary zone electrophoresis (CZE) have been used for betalain and anthocyanins analyses. Regarding the first, the use of pyridine and formic acid, pyridine and citric acid or acetic acid as solvents or in cellulose, are common and reliable methods for betacyanin determinations in red beet (Powrie and Fennema 1963; Piatelli and Minale 1964; von Elbe et al. 1972). More recently, a CZE method was developed for the separation and determination of betanin and isobetanin in different extracts of *Beta vulgaris* (Stuppner and Egger 1996). The method was used with a fused-silica capillary at 15 °C and at a constant voltage (22 kV), using citrate-phosphate buffer (10 mM, pH 6.0) as running electrolyte; the detection wavelengths were 538 and 477 nm. It was found that the results for betalain quantification based on CZE analyses were in close agreement with those obtained by HPLC. Capillary zone electrophoresis required longer analysis time, which could be dramatically shortened by focusing the analysis only on the separation of the two major red pigments, leaving unseparated the agylcones betanidin and isobetanidin, and also the yellow pigments. Capillary zone electrophoresis was also used for the separation of anthocyanin pigments from strawberry and elderberry fruit (Bridle 1996; Bridle and Garcia-Viguera 1997). On a comparative basis, HPLC provided more advantages than CZE methods. CZE lacked flexibility and requires higher sample concentration. Nevertheless, CZE has the potential to optimise separations by varying other parameters (e.g., temperature, voltage, electrolyte concentration). It also allows the use of additives (complexing agents, organic modifiers, and micelles), particularly if methodology for working with strong acidic buffers becomes available. Bicard et al. (1999) used CZE in acidic media to analyse natural anthocyanins from black glutinous rice. Nevertheless, the use of CZE under acidic conditions can significantly improve peak resolution and detection limits, but it does not allow the separation of complex samples that can be achieved with HPLC (Mazza et al. 2004).

5.3.2 Non-chromatographic Techniques

The most widely employed techniques for determination of pigments in fruit and vegetables are based on ion or gas chromatography, which may be coupled to diverse detection systems. Nevertheless, these methods require expensive instrumentation and long analysis time, making them unsuitable for on-site routine measurements. To overcome these drawbacks, the use of non-chromatographic methodologies has been reported as alternative. Most of them exploit optical properties of the product, such as reflectance, transmittance, absorbance, scatter of light or fluorescence. Although the majority of these methods maintain the product intact, being non-invasive or non-destructive, some invasive methods have also been applied.

Non-invasive Methods

Extensive research has been performed on developing non-destructive methodologies to evaluate maturity of fruit and vegetables based on the assessment of their physical, mechanical and optical properties. Optical methods have been playing an important role in this field, due to the miniaturization, portability, robustness, low cost and high accuracy offered by the commercially available spectrophotometer devices. Among these methods, visible and near infrared (NIR) spectroscopy, time-resolved reflectance spectroscopy, hyperspectral backscattering imaging and laser-induced light backscattering were employed for non-destructive measurement of quality in fruit and vegetables. Most of these techniques were recently reviewed by Nicolaï et al. (2007) as NIR spectroscopy techniques. The product is irradiated with NIR radiation, and the reflected or transmitted radiation is measured. While the radiation penetrates the product, its spectral characteristics change through wavelength-dependent scattering and absorption. This change depends on the chemical composition of the product, as well as on its light-scattering properties, which are related to the microstructure.

Visible spectra obtained non-destructively from the fruit samples are frequently analyzed in terms of color using various color spaces, thus providing information on the pigment contents (Zude et al. 2011). Interferences caused by similarity of absorption wavelengths of different pigments can be minimized by using specific indices or whole spectra analyses. Several indices have been established in feasible applications, mainly for chlorophyll analysis. The background color of fruit is provided by the chlorophyll content of the skin and associated tissue, and the decrease in the pigment content during maturation may be considered an indicator of the stage of fruit development. These principles have been applied to physiological evaluations of the fruit maturity, such as the chlorophyll ripeness of apples (Merzlyak et al. 1999; Zude 2003; Peirs et al. 2005; Solovchenko et al. 2005; Herold et al. 2005; Infante et al. 2011; Venturello et al. 2012; Bertone et al. 2012), peaches (Ziosi et al. 2008; Herrero-Langreo et al. 2012) and mangoes (Betemps et al. 2011). Chlorophyll *a*, carotenoids and anthocyanins were determined in durians (Timkhum and Terdwongworakul 2012) and anthocyanins were assessed in cherries (Zude et al. 2011). This technique can also be used for nutritional assessments, such as determination of the content of native carotenoids in carrots (Zude et al. 2007) and lycopene in tomato (Fernandez-Ruiz et al. 2010).

Reflectance spectra in the visible and near infra-red range were used to evaluate pigment changes during leaf senescence and ripening of maple, chestnut, potato, and coleus leaves, and lemon and apple fruit (Merzlyak et al. 1999). An increase of reflectance between 550 and 740 nm accompanied senescence-induced degradation of chlorophyll, whereas in the range of 400–500 nm it remained low, due to retention of carotenoids. It was found that both leaf senescence and fruit ripening affect the difference between reflectance around 670 and 500 nm, depending on pigment composition.

Biospeckle is another optical technique based on the illumination of the product by coherent laser light. The backscattered light creates a specific speckle pattern in the observation plane. Speckle pattern is maintained stable in time, if the sample does not show any physiological activity (Xu et al. 1995). Although the knowledge about biospeckle in relation to fruit and vegetables is still limited, studies suggest that biospeckle activity changes with the developmental stage of the biological material (Zdunek and Herppich 2012). This technique was applied to the determination of chlorophyll content in tomato (Romero et al. 2009) and apples (Zdunek and Herppich 2012) using a red laser, which light is absorbed by chlorophyll. It was concluded that biospeckle measurements were not feasible to estimate chlorophyll contents in apples. Additionally, the results indicated that variation in biospeckle activity did not fully reflect true changes in biological activity if chlorophyll contents changed.

Multispectral fluorescence is another non-invasive optical method widely used to assess quality indices of fruit and vegetables during ripening, by exploiting the ability of pigments to produce fluorescence after excitation by a light source. The properties of excitation light source, such as type of signal (continuous or modulated), and intensity and spectral characteristics, determine the fluorescence information that can be obtained. Excitation in different wavelength ranges may provide information from different plant tissues (Vogelmann 1993). In the visible range, and according to Lenk et al. (2007), blue light is absorbed with high efficiency by carotenoids and chlorophylls and red light is absorbed only by chlorophylls. Fluorescence-based methods were developed to determine non-invasively the contents of chlorophylls in diverse fruit and vegetables such as apples (Moshou et al. 2003; Bodria et al. 2004; Greer 2005; Hagen et al. 2007; Solovchenko et al. 2010), cherries (Linke et al. 2010), kale (Hagen et al. 2009), grapes (Ramin et al. 2008), papaya (Bron et al. 2004) and peaches (Bodria et al. 2004). Methods for determination of anthocyanins in grapes were also reported (Ghozlen et al. 2010; Baluja et al. 2012). Multi-pigment analyses were performed on grapes (Cerovic et al. 2008), tomatoes (Lai et al. 2007) and apples (Hagen et al. 2006; Merzlyak et al. 2008; Betemps et al. 2012). Some of these methods were based on the screening of fruit chlorophyll fluorescence to determine the contents of other pigments (Cerovic et al. 2008). Merzlyak et al. (2008) developed a simple model that allowed the simulation of chlorophyll fluorescence excitation spectra in the visible range and a quantitative evaluation of competitive absorption by anthocyanins, carotenoids and flavonols. In fruit with low-to-moderate pigment content, those components play the role of internal traps (insofar as they compete with chlorophylls for the absorption of incident light in specific spectral bands), affecting thereby the shape of the chlorophyll fluorescence excitation spectrum.

Invasive Methods

Costache et al. (2012) applied spectrophotometry to simultaneously determine chlorophyll and carotene content in different vegetables, testing three organic solvents to obtain the best extraction solution. The literature presents trichromatic and monochromatic methods to spectrophotometrically determine these pigments, differing by the used wavelengths. Since chlorophyll pigments have a broad absorption band from blue to red, and co-extracted carotenoids have the maximum absorption in the blue band, they chose the trichromatic method. This method is based on the measurement of the absorbance at three wavelengths for each type of matrix, after extraction of carotenoid and chlorophyll pigments in methanol, diethyl ether and acetone. After analyses of 17 samples of cherry tomatoes, peppers and cucumbers, chlorophylls *a*, *b* and *c* were the main pigments found in these vegetables.

Spectrophotometry was also applied to chlorophyll determination in fresh-cut Swiss chard leaves after extraction with acetone (Ferrante et al. 2008; Kasim and Kasim 2012). Chlorophyll content was measured at 645 and 663 nm and the total chlorophyll content was obtained by the sum of chlorophyll *a* and *b* contents (Kasim and Kasim 2012). Ishak et al. (2005) evaluated some physical and chemical properties of ambarella fruit at three different stages of maturity. Chlorophyll content was calculated by substituting the results from the spectrophotometric readings at 660.0nm ($A_{660.0}$) and 642.5 nm ($A_{642.5}$) into the following equations:

total chlorophyll= $7.12\ A_{660.0} + 16.8 A_{642.5}$ eq. (5.1)

chlorophyll *a* = $9.93 A_{660.0} + 0.777 A_{642.5}$ eq. (5.2)

chlorophyll *b*= $17.6 A_{660.0} + 2.81 A_{642.5}$ eq. (5.3)

Biehler et al. (2010) compared three spectrophotometric methods, all based on rapid extraction, to determine the total amount of carotenoids present in fruit and vegetables, either with or without chlorophyll. The authors conclude that the typical overestimation by spectrophotometric methods due to minor compounds and degradation products is somewhat balanced by carotenoid losses due to saponification, resulting in a close estimation of the carotenoid content obtained by HPLC.

According to Moshou et al. (2005), the chlorophyll fluorescence in apples follows the same course as in plant leaves, but with less intensity because of a lower chlorophyll content of the apple fruit. In sorting fruit, maintaining the ambient conditions constant while measuring the changes in fluorescence can be linked to the ageing and ripening, causing decrease of firmness, chlorophyll loss and loss of photosynthetic activity per unit chlorophyll.

Infrared spectroscopy gives information on the kind of bonds and atoms in the analyzed compound, while NMR on protons and carbon-13 permits the assigning of these atoms to a certain structure (Delgado-Vargas et al. 2000). In fact, NMR spectroscopy has been demonstrated to be a powerful technique for identifying and determining the structural properties of chlorophyll derivatives and carotenoids (Valverde and This 2008). These authors presented a study of the direct identification and quantification of the main photosynthetic pigments in green bean crude extracts using NMR methods. The method enables the attainment of more information on chlorophyll derivatives (allomers and epimers) than UV-Vis spectroscopy, without prior chromatographic separation.

The difficulty in separating single carotenoids using spectrophotometric transmittance readings of fruit extracts caused by coinciding absorption bands of the various carotenoids and chlorophylls present in the solution was overcome by Pflanz and Zude (2008). They developed an iteratively applied linear regression based on spectral profiles of pigment standards. The iterative approach was validated by dilution series of pigments and compared with commonly used equation systems. The method was applied to tomatoes and high coefficients of determination and low measuring uncertainties were found for chlorophyll a and b, as well as for carotenoids, such as lycopene, β-carotene, and lutein.

5.4 Conclusion

Research on pigments of fruit and vegetables is challenging, involving many analytical methods and a deep knowledge in several fields. The identification and quantification of natural pigments are important for identifying target foods, whose consumption helps in health-promotion and disease prevention, presenting an added value in the functional perspective.

Detection of pigments in post-harvesting of fruit and vegetables, and their correlation with the ripeness and senescence is also an area of interest.

The assessment of correlations between pigment compounds and bioactivity is still an area with potential for exploitation. This information may be integrated and analyzed using a chemometric approach to correlate the foods, bioactive compounds profile with the antioxidant activity and thus to predict antioxidant activity of fruit and vegetables.

For all the above-mentioned purposes, it is important to have the most powerful methodologies at our disposal to analyze the pigment composition of samples. Non-invasive optical methods, such as visible and near-infrared spectroscopies, and NMR spectroscopies, can be useful technologies, which, in the past, were restricted to basic research.

Acknowledgements

This work was supported by the Portuguese national agency FCT through the project PEst-OE/EQB/LA0016/2011.

References

Ade-Omowaye, B.I.O., Angersbach, A., Taiwo, K.A. and Knorr, D. 2001. Use of pulsed electric field pre-treatment to improve dehydration characteristics of plant based foods. Trends Food Sci. Tech. 2: 285–295.

Aman, R., Schieber, A. and Carle, R. 2005a. Effects of heating and illumination on trans-cis isomerization and degradation of β-carotene and lutein in isolated spinach chloroplasts. J. Agric. Food Chem. 53: 9512–9518.

Aman, R., Biehl, J., Carle, R., Conrad, J., Beifuss, U. and Schieber, A. 2005b. Application of HPLC coupled with DAD, APcI-MS and NMR to the analysis of lutein and zeaxanthin stereoisomers in thermally processed vegetables. Food Chem. 92: 753–763.

Ambrogi, A., Cardarelli, R. and Eggers, R. 2002. Fractional extraction of paprika using supercritical carbon dioxide and on-line determination of carotenoids. J. Food Sci. 67: 3236–3241.

Andersen, Ø.M., Fossen, T., Torskangerpoll, K., Fossen, A. and Hauge, U. 2004. Anthocyanin from strawberry (*Fragaria ananassa*) with the novel aglycone, 5-carboxypyranopelargonidin. Phytochemistry. 65: 405–410.

Anderson, D.W., Gueffroy, D.E., Webb, A.D. and Kepner, R.E. 1970. Identification of acetic acid as an acylating agent of anthocyanin pigments in grapes. Phytochemistry. 9: 1579–1583.

Arapitsas, P., Sjöberg, P.J.R. and Turner, C. 2008. Characterisation of anthocyanins in red cabbage using high resolution liquid chromatography coupled with photodiode array detection and electrospray ionization-linear ion trap mass spectrometry. Food Chem. 109: 219–226.

Aruna, G., Mamatha, B.S. and Baskaran, V. 2009. Lutein content of selected Indian vegetables and vegetable oils determined by HPLC. J. Food Compos. Anal. 22: 632–636.

Azeredo, H.M.C. 2009. Betalains: properties, sources, applications, and stability—a review. Int. J. Food Sci. Tech. 44: 2365–2376.

Baluja, J., Diago, M.P., Goovaerts, P. and Tardaguila, J. 2012. Spatio-temporal dynamics of grape anthocyanin accumulation in a Tempranillo vineyard monitored by proximal sensing. Aust. J. Grape Wine R. 18: 173–183.

Barba, A.I.O., Hurtado, M.C., Mata, M.C.S., Ruiz, V.F. and de Tejada, M.L.S. 2006. Application of a UV-Vis detection-HPLC method for a rapid determination of lycopene and beta-carotene in vegetables. Food Chem. 95: 328–336.

Barth, M.M., Zhou, C., Kute, K.M. and Rosenthal, G.A. 1995. Determination of optimum conditions for supercriticalfluid extraction of carotenoids from carrot (*Daucus carota* L.) tissue. J. Food Chem. 43: 2876–2878.

Bartley, G.E. and Scolnik, P.A. 1995. Plant carotenoids: Pigments for photoprotection, visual attraction, and human health. Plant Cell. 7: 1027–1038.

Bayindirli, A., Yildiz, F. and Özilgen, M. 1988. Modeling of sequential batch ultrafiltration of red beet extract. J. Food Sci. 53: 1418–1422.
Bell, C.M., Sander, L.C., Fetzer, J.C. and Wise, S.A. 1996. Synthesis and characterization of extended length alkyl stationary phases for liquid chromatography with application to the separation of carotenoid isomers. J. Chromatogr. A. 753: 37–45.
Bertone, E., Venturello, A., Leardi, R. and Geobaldo, F. 2012. Prediction of the optimum harvest time of 'Scarlet' apples using DR-UV-Vis and NIR spectroscopy. Postharvest Biol. Tech. 69: 15–23.
Betemps, D.L., Fachinello, J.C. and Galarca, S.P. 2011. Visible spectroscopy and near infrared (Vis/NIR), in assessing the quality of mangoes Tommy Atkins. Rev. Bras. Frutic. 33: 306–313.
Betemps, D.L., Fachinello, J.C., Galarca, S.P., Portela, N.M., Remorini, D., Massai, R. and Agati, G. 2012. Non-destructive evaluation of ripening and quality traits in apples using a multiparametric fluorescence sensor. J. Sci. Food Agr. 92: 1855–1864.
Bicard, V., Fougerousse, A. and Brouillard, R. 1999. Analysis of natural anthocyanins by capillary zone electrophoresis in acidic media. J. Liquid Chromatogr. Rel. Tech. 22: 541–550.
Biehler, E., Mayer, F., Hoffmann, L., Krause, E. and Bohn, T. 2010. Comparison of 3 spectrophotometric methods for carotenoid determination in frequently consumed fruits and vegetables. J. Food Sci. 75: C55–C61.
Bilyk, A. 1981. Thin-layer chromatographic separation of beet pigments. J. Food Sci. 46: 298–299.
Bodria, L., Fiala, M., Guidetti, R. and Oberti, R. 2004. Optical techniques to estimate the ripeness of red-pigmented fruits. T. ASAE 47: 815–820.
Bohn, T. and Walczyk, T. 2004. Determination of chlorophyll in plant samples by liquid chromatography using zinc-phthalocyanine as an internal standard. J. Chromatogr. A. 1024: 123–128.
Bononi, M., Andreoli, G., Granelli, G., Eccher, T. and Tateo, F. 2006. 'Cyanidin volumetric index' and 'chromaticity coordinates ratio' to characterize red raspberry (*Rubus idaeus*). Int. J. Food Sci. Nutr. 57: 369–375.
Bordonaba, J.G., Crespo, P. and Terry, L.A. 2011. A new acetonitrile-free mobile phase for HPLC-DAD determination of individual anthocyanins in blackcurrant and strawberry fruits: A comparison and validation study. Food Chem. 129: 1265–1273.
Bridle, P. 1996. Analysis of anthocyanins by capillary zone electrophoresis. J. Liquid Chromatogr. Rel. Tech. 19: 537–545.
Bridle, P. and Garcia-Viguera, C. 1997. Analysis of anthocyanins in strawberries and elderberries. A comparison of capillary zone electrophoresis and HPLC. Food Chem. 59: 299–304.
Bron, I.U., Ribeiro, R.V., Azzolini, M., Jacomino, A.P. and Machado, E.C. 2004. Chlorophyll fluorescence as a tool to evaluate the ripening of 'Golden' papaya fruit. Postharvest Biol. Tech. 33: 163–173.
Brown, P.N. and Shipley, P.R. 2011. Determination of anthocyanins in cranberry fruit and cranberry fruit products by High-Performance Liquid Chromatography with ultraviolet detection: single-laboratory validation. J. AOAC Int. 94: 459–466.
Can, N.O., Arli, G. and Atkosar, Z. 2012. Rapid determination of free anthocyanins in foodstuffs using high performance liquid chromatography. Food Chem. 130: 1082–1089.
Cano, M.P. and Marín, M.A. 1992. Pigment composition and color of frozen and canned kiwi fruit slices. J. Agr. Food Chem. 40: 2141–2146.
Cao, S., Liu, L., Pan, S., Lu, Q. and Xu, X. 2009. A comparison of two determination methods for studying degradation kinetics of the major anthocyanins from blood orange. J. Agr. Chem. 57: 245–249.
Castellar, M.R., Obón, J.M. and Fernández-López, J.A. 2006. The isolation and properties of a concentrated red-purple betacyanin food colourant from Opuntia stricta fruits. J. Sci. Food Agr. 86: 122–128.

Cerovic, Z.G., Moise, N., Agati, G., Latouche, G., Ghozlen, N.B. and Meyer, S. 2008. New portable optical sensors for the assessment of winegrape phenolic maturity based on berry fluorescence. J. Food Compos. Anal. 21: 650–654.
Chalermchat, Y., Dejmek, P. and Fincan, M. 2004. Pulsed electric field treatment for solid-liquid extraction of red beetroot pigment: mathematical modelling of mass transfer. J. Food Eng. 64: 229–236.
Chester, T.L., Pinkston, J.D. and Raynie, D.E. 1992. Supercritical fluid chromatography and extraction. Anal. Chem. 64: 153R–170R.
Chernomorsky, S., Rancourt, R., Sahai, D. and Poretz, R. 1997. Evaluation of commercial chlorophyllin copper complex preparations by liquid chromatography with photodiode array detection. J. AOAC Int. 80: 433–435.
Choung, M.G., Choi, B.R., An, Y.N., Chu, Y.H. and Cho, Y.S. 2003. Anthocyanin profile of Korean cultivated kidney bean (*Phaseolus vulgaris* L.). J. Agr. Food Chem. 51: 7040–7043.
Clement, J.S. and Mabry, T.J. 1996. Pigment evolution in the Caryophyllales: A systematic overview. Bot. Acta 109: 360–367.
Costache, M.A., Campeanu, G. and Neata, G. 2012. Studies concerning the extraction of chlorophyll and total carotenoids from vegetables. Rom. Biotech. Lett. 17: 7702–7708.
Cucu, T., Huvaere, K., Van den Bergh, M.A., Vinkx, C. and Van Loco, J. 2012. A simple and fast HPLC method to determine lycopene in foods. Food Anal. Method. 5: 1221–1228.
Dachtler, M., Glaser, T., Kohler, K. and Albert, K. 2001. Combined HPLC-MS and HPLC-NMR on-line coupling for the separation and determination of lutein and zeaxanthin stereoisomers in spinach and in retina. Anal. Chem. 73: 667–674.
Daurade-LeVagueresse, M.H. and Bounias, M. 1991. Separation, quantification, spectral properties and stability of photosynthetic pigments on CN-coated HPTLC plates. Chromatographia. 31: 5–10.
de Rosso, V.V., Hillebrand, S., Montilla, E.C., Bobbio, F.O., Winterhalter, P. and Mercadante, A.Z. 2008. Determination of anthocyanins from acerola (*Malpighia emarginata* DC.) and açaí (*Euterpe oleracea* Mart.) by HPLC–PDA–MS/MS. J. Food Compos. Anal. 21: 291–299.
Delgado-Vargas, F., Jiménez, A.R. and Paredes-López, O. 2000. Natural pigments: carotenoids, anthocyanins, and betalains—characteristics, biosynthesis, processing, and stability. Crit. Rev. Food Sci. 40: 173–289.
Deschene, A., Paliyath, G., Lougheed, E.C., Dumbroff, E.B. and Thompson, J.E. 1991. Membrane deterioration during postharvest senescence of broccoli florets: modulation by temperature and controlled atmosphere storage. Postharvest Biol. Tech. 1: 19–31.
Dias, M.G., Camões, M.F.G.F.C. and Oliveira, L. 2008. Uncertainty estimation and in-house method validation of HPLC analysis of carotenoids for food composition data production. Food Chem. 109: 815–824.
Dias, M.G., Camões, M.F.G.F.C. and Oliveira, L. 2009. Carotenoids in traditional Portuguese fruits and vegetables. Food Chem. 113: 808–815.
Dugo, P., Mondello, L., Errante, G., Zappia, G. and Dugo, G. 2001. Identification of anthocyanins in berries by narrow-bore high-performance liquid chromatography with electrospray ionization detection. J. Agr. Food Chem. 49: 3987–3992.
Edelenbos, M., Christensen, L.P. and Grevse, K. 2001. HPLC determination of chlorophyll and carotenoid pigments in processed green pea cultivars (*Pisum sativum* L.). J. Agr. Food Chem. 49: 4768–4774.
Eder, R. 2000. Pigments. pp. 825–880. *In*: Nollet, L.M.L. [ed.]. Food Analysis by HPLC.Marcel Dekker, New York, USA.
Emenhiser, C., Sander, L.C. and Schwartz, S.J. 1995. Capability of a polymeric C-30 stationary-phase to resolve cis-trans carotenoid isomers in reversed-phase liquid chromatography. J. Chromatogr. A. 707: 205–216.
Feltl, L., Pacáková, V., Štulík, K. and Volka, K. 2005. Reliability of carotenoid analyses: a review. Curr. Anal. Chem. 1: 93–102.

Fernández-García, E., Carvajal-Lérida, I., Jarén-Galán, M., Garrido-Fernández, J., Pérez-Gálvez, A. and Hornero-Méndez, D. 2012. Carotenoids bioavailability from foods: From plant pigments to eficient biological activities. Food Res. Int. 46: 438–450.

Fernandez-Jaramillo, A.A., Duarte-Galvan, C., Contreras-Medina, L.M., Torres-Pacheco, I., Romero-Troncoso, R.J., Guevara-Gonzalez, R.G. and Millan-Almaraz, J.R. 2012. Instrumentation in developing chlorophyll fluorescence biosensing: a review. Sensors 12: 11853–11869.

Fernández-León, M.F., Lozano, M., Ayuso, M.C., Fernández-León, A.M. and González-Gómez, D. 2010. Fast and accurate alternative UV-chemometric method for the determination of chlorophyll A and B in broccoli (*Brassica oleracea* Italica) and cabbage (*Brassica oleracea* Sabauda) plants. J. Food Compos. Anal. 23: 809–813.

Fernández-Ruiz, V., Torrecilla, J.S., Camara, M., Mata, M.C.S. and Shoemaker, C. 2010. Radial basis network analysis of color parameters to estimate lycopene content on tomato fruits. Talanta. 83: 9–13

Ferrante, A., Incrocci, L. and Serra, G. 2008. Quality changes during storage of fresh-cut or intact Swiss chard leafy vegetables. J. Food Agric. Environ. 6: 60–62.

Fincan, M., De Vito, F. and Dejmek, P. 2004. Pulsed electric field treatment for solid-liquid extraction of red beetroot pigment. J. Food Eng. 64: 381–388.

Fleshman, M.K., Lester, G.E., Riedl, K.M., Kopec, R.E., Narayanasamy, S., Jr. Curley, R.W., Schwartz, S.J. and Harrison, E.H. 2011. Carotene and novel apocarotenoid concentrations in orange-fleshed *Cucumis melo* melons: determinations of β-carotene bioaccessibility and bioavailability. J. Agr. Food Chem. 59: 4448–4454.

Garcia-Viguera, C., Zafrilla, P. and Tomás-Barberán, F.A. 1998. The use of acetone as an extraction solvent for anthocyanins from strawberry fruit. Phytochem. Analysis 9: 274–277.

Garzón, G.A., Riedl, K.M. and Schwartz, S.J. 2009. Determination of anthocyanins, total phenolic content, and antioxidant activity in Andes berry (*Rubus glaucus* Benth). J. Food Sci. 74: C227–C232.

Garzón, G.A., Narvaez-Cuenca, C.E., Kopec, R.E., Barry, A.M., Riedl, K.M. and Schwartz, S.J. 2012. Determination of carotenoids, total phenolic content, and antioxidant activity of Araza (*Eugenia stipitata* McVaugh), an amazonian fruit. J. Agr. Food Chem. 60: 4709–4717.

Gentili, A. and Caretti, F. 2011. Evaluation of a method based on liquid chromatography–diode array detector–tandem mass spectrometry for a rapid and comprehensive characterization of the fat-soluble vitamin and carotenoid profile of selected plant foods. J. Chromatogr. A. 1218: 684–697.

Ghozlen, N.B., Cerovic, Z.G., Germain, C., Toutain, S. and Latouche, G. 2010. Non-destructive optical monitoring of grape maturation by proximal sensing. Sensors. 10: 10040–10068.

Girod, P.A. and Zryd, J.P. 1987. Isolation and culture of betaxanthins and betacyanins producing cells of red beet (*Beta vulgaris* L.). Experientia. 43: 660–661.

Giuffrida, D., Torre, G., Dugo, P. and Dugo, G. 2013. Determination of the carotenoid profile in peach fruits, juice and jam. Fruits. 68: 39–44.

Giusti, M.M., Rodríguez-Saona, L.E., Griffin, D. and Wrolstad, R.E. 1999. Electrospray and Tandem mass spectroscopy as tools for anthocyanin characterization. J. Agric. Food Chem. 47: 4657–4664.

Glaser, T., Lienau, A., Zeeb, D., Krucker, M., Dachtler, M. and Albert, K. 2003. Qualitative and quantitative determination of carotenoid stereoisomers in a variety of spinach samples by use of MSPD before HPLC-UV, HPLC-APCI-MS, and HPLC-NMR on-line coupling. Chromatographia. 57: S19–S25.

Gnayfeed, M.H., Daood, H.G., Illes, V. and Biacs, P.A. 2001. Supercritical CO_2 and subcritical propane extraction of pungent paprika and quantification of carotenoids, tocopherols, and capsaicinoids. J. Agricult. Food Chem. 49: 2761–2766.

Greer, D.H. 2005. Non-destructive chlorophyll fluorescence and colour measurements of 'Braeburn' and 'Royal Gala' apple (*Malus domestica*) fruit development throughout the growing season. New Zeal. J. Crop Hort. 33: 413–421.

Hagen, S.F., Borge, G.I.A., Bengtsson, G.B., Bilger, W., Berge, A., Haffner, K. and Solhaug, K.A. 2007. Phenolic contents and other health and sensory related properties of apple fruit (*Malus domestica* Borkh., cv. Aroma): Effect of postharvest UV-B irradiation. Postharvest Biol. Tech. 45: 1–10.

Hagen, S.F., Borge, G.I.A., Solhaug, K.A. and Bengtsson, G.B. 2009. Effect of cold storage and harvest date on bioactive compounds in curly kale (*Brassica oleracea* L. var. acephala). Postharvest Biol. Tech. 51: 36–42.

Hagen, S.F., Solhaug, K.A., Bengtsson, G.B., Borge, G.I.A. and Bilger, W. 2006. Chlorophyll fluorescence as a tool for non-destructive estimation of anthocyanins and total flavonoids in apples. Postharvest Biol. Tech. 41: 156–163.

Hawthorne, S.B. 1990. Analytical-scale supercritical fluid extraction. Anal. Chem. 62: 633A–642A.

He, X.-G., Lin, L.-Z., Lian, L.-Z. and Lindenmaier, M. 1998. Liquid chromatography electrospray mass spectrometric analysis of curcuminoids and sesquiterpenoids in turmeric (*Curcuma longa*). J. Chromatogr. A. 818: 127–132.

Henning, W. 1983. Detection and evaluation of the addition of betanin to foodstuffs. Deut. Lebensm-Rundsch. 79: 407–410.

Herbach, K.M., Stintzing, F.C., Elss, S., Preston, C., Schreier, P. and Carle, R. 2006. Isotope ratio mass spectrometrical analysis of betanin and isobetanin isolates for authenticity evaluation of purple pitaya-based products. Food Chem. 99: 204–209.

Herold, B., Truppel, I., Zude, M. and Geyer, M. 2005. Spectral measurements on 'Elstar' apples during fruit development on the tree. Biosyst. Eng. 91: 173–182.

Herrero-Langreo, A., Fernandez-Ahumada, E., Roger, J.-M., Palagos, B. and Lleo, L. 2012. Combination of optical and non-destructive mechanical techniques for the measurement of maturity in peach. J. Food Eng. 108: 150–157.

Hong, V. and Wrolstad, R.E. 1990. Use of HPLC separation photodiode array detection for characterization of anthocyanins. J. Agr. Food Chem. 38: 708–715.

Hornero-Méndez, D. and Minguez-Mosquera, M.I. 1998. Isolation and identification of the carotenoid capsolutein from *Capsicum annuum* as cucurbitaxanthin A. J. Agric. Food Chem. 46: 4087–4090.

Hornero-Méndez, D. and Mínguez-Mosquera, M.I. 2000. Xanthophyll esterification accompanying carotenoid overaccumulation in chromoplast of *Capsicum annuum* ripening fruits is a constitutive process and useful for ripeness index. J. Agric. Food Chem. 48: 1617–1622.

Ieri, F., Pinelli, P. and Romani, A. 2012. Simultaneous determination of anthocyanins, coumarins and phenolic acids in fruits, kernels and liqueur of *Prunus mahaleb* L. Food Chem. 135: 2157–2162.

Infante, R., Rubio, P., Contador, L., Noferini, M. and Costa, G. 2011. Determination of harvest maturity of D'Agen plums using the chlorophyll absorbance index. Cien. Investig. Agrar. 38: 199–203.

Isaksen, M. and Francis, G.W. 1986. Reversed-phase thin-layer chromatography of carotenoids. J. Chromatogr. 355: 358–362.

Ishak, S.A., Ismail, N., Noor, M.A.M. and Ahmad, H. 2005. J. Food Compos. Anal. 18: 819–827.

Jarén-Galán, M., Nienaber, U. and Schwartz, S.J. 1999. Paprika (*Capsicum annuum*) oleoresin extraction with supercritical carbon dioxide. J. Agric. Food Chem. 47: 3558–3564.

Jiménez, C.D.C., Flores, C.S., He, J., Tian, Q., Schwartz, S.J. and Giusti, M.M. 2011. Characterisation and preliminary bioactivity determination of *Berberis boliviana* Lechler fruit anthocyanins. Food Chem. 128: 717–724.

Jubert, C. and Bailey, G. 2007. Isolation of chlorophylls *a* and *b* from spinach by counter-current chromatography. J. Chromatogr. A. 1140: 95–100.

Kanasawud, P. and Crouzet, J.C. 1990. Mechanism of formation of volatile compounds by thermal-degradation of carotenoids in aqueous-medium.1. beta-carotene degradation. J. Agric. Food Chem. 38: 237–243.

Kao, T.H., Chen, C.J. and Chen, B.H. 2011. An improved high performance liquid chromatography–photodiode array detection–atmospheric pressure chemical ionization–mass spectrometry method for determination of chlorophylls and their derivatives in freeze-dried and hot-air-dried *Rhinacanthus nasutus* (L.) Kurz. Talanta. 86: 349–355.

Kapadia, G.J., Azuine, M.A., Sridha, R., Okuda, Y., Tsuruta, A., Ichiishi, E., Mukainake, T., Takasaki, M., Konoshima, T., Nishino, H. and Tokuda, H. 2003. Chemoprevention of DMBA-induced UV-B promoted, NOR-1-induced TPA promoted skin carcinogenesis, and DEN-induced phenobarbital promoted liver tumors in mice by extract of beetroot. Pharmacol. Res. 47: 141–148.

Kapadia, G.J., Tokuda, H., Konoshima, T. and Nishino, H. 1996. Chemoprevention of lung and skin cancer by *Beta vulgaris* (beet) root extract. Cancer Lett. 100: 211–214.

Kasim, M.U. and Kasim, R. 2012. Color changes of fresh-cut Swiss chard leaves stored at different light intensity. Am. J. Food Technol. 7: 13–21.

Khoo, H., Prasad, K.N., Kong, K., Jiang, Y. and Ismail, A. 2011. Carotenoids and their isomers: color pigments in fruits and vegetables. Molecules 16: 1710–1738.

Kidmose, U., Yang, R.-Y., Thilsted, S.H., Christensen, L.P. and Brandt, K. 2006. Content of carotenoids in commonly consumed Asian vegetables and stability and extractability during frying. J. Food Compos. Anal. 19: 562–571.

Konczak-Islam, I., Okunob, S., Yoshimoto, M. and Yamakawa, O. 2003. Composition of phenolics and anthocyanins in a sweet potato cell suspension culture. Biochem. Eng. J. 14: 155–161.

Kong, J., Chia, L., Goh, N., Chia, T. and Brouillard, R. 2003. Analysis and biological activities of anthocyanins. Phytochemistry. 64: 923–933.

Konings, E.J.M. and Roomans, H.H.S.. 1996. Evaluation and validation of an LC method for the analysis of carotenoids in vegetables and fruit. Food Chem. 59: 599–603.

Kotíková, Z., Hejtmánková, A. and Lachman. J. 2009. Determination of the influence of variety and level of maturity on the content and development of carotenoids in tomatoes. Czech J. Food Sci. 27: S200–S203.

Kotíková, Z., Lachmanm, J., Hejtmánková, A. and Hejtmánková, K. 2011. Determination of antioxidant activity and antioxidant content in tomato varieties and evaluation of mutual interactions between antioxidants. Lebensm-Wiss Technol. 44: 1703–1710.

Lacker, T., Strohschein, S. and Albert, K. 1998. Separation and identification of various carotenoids by C-30 reversed-phase high-performance liquid chromatography coupled to UV and atmospheric pressure chemical ionization mass spectrometric detection. J. Chromatogr. A. 854: 37–44.

Lai, A., Santangelo, E., Soressi, G.P. and Fantoni, R. 2007. Analysis of the main secondary metabolites produced in tomato (*Lycopersicon esculentum*, Mill.) epicarp tissue during fruit ripening using fluorescence techniques. Postharvest Biol. Tech. 43: 335–342.

Lancaster, J.E., Lister, C.E., Reay, P.F. and Triggs, C.M. 1997. Influence of pigment composition on skin color in a wide range of fruit and vegetables. J. Am. Soc. Hortic. Sci. 122: 594–598.

Lee, Y.N. and Wiley, R.C. 1981. Betalaine yield from a continuous solid-liquid extraction system as influenced by raw product, postharvest and processing variables. J. Food Sci. 46: 421–424.

Lee, M.T. and Chen, B.H. 2001. Separation of lycopene and its *cis* isomers by liquid chromatography. Chromatographia. 54: 613–617.

Lee, Y.N., Wiley, R.C., Sheu, M.J. and Schlimme, D.V. 1982. Purification and concentration of betalaines by ultrafiltration and reverse osmosis. J. Food Sci. 47: 465–471.

Lenk, S., Chaerle, L., Pfündel, E.E., Langsdorf, G., Hagenbeek, D., Lichtenthaler, H.K., Van Der Straeten, D. and Buschmann, C. 2007. Multispectral fluorescence and reflectance imaging at the leaf level and its possible applications. J. Exp. Bot. 58: 807–814.

Lima, A.J.B., Corrêa, A.D., Saczk, A.A., Martins, M.P. and Castilho, R.O. 2011. Anthocyanins, pigment stability and antioxidant activity in jabuticaba [*Myrciaria cauliflora* (Mart.) O. Berg]. Rev. Bras. Frutic. 33: 877–887.
Linke, M., Herppich, W.B. and Geyer, M. 2010. Green peduncles may indicate postharvest freshness of sweet cherries. Postharvest Biol. Tech. 58: 135–141.
Loh, C.H., Inbaraj, B.S., Liu, M.H. and Chen, B.H. 2012. Determination of chlorophylls in *Taraxacum formosanum* by high-performance liquid chromatography-diode array detection-mass spectrometry and preparation by column chromatography. J. Agr. Food Chem. 60: 6108–6115.
Longo, L. and Vasapollo, G. 2005. Determination of anthocyanins in *Ruscus aculeatus* L. berries. J. Agric. Food Chem. 53: 475–479.
Losciale, P., Hendrickson, L., Grappadelli, L.C. and Chow, W.S. 2011. Quenching partitioning through light-modulated chlorophyll fluorescence: a quantitative analysis to assess the fate of the absorbed light in the field. Environ. Exp. Bot. 73: 73–79.
Lucini, L., Pellizzoni, M., Baffi, C. and Molinari, G.P. 2012. Rapid determination of lycopene and β-carotene in tomato by liquid chromatography/electrospray tandem mass spectrometry. J. Sci. Food Agric. 92: 1297–1303.
Mabry, T.J. 1980. Betalains. pp. 513–533. *In:* E.A. Bell and B.V. Charlwood [eds.]. Encyclopedia of Plant Physiology, Vol. 8, Secondary Plant Products. Springer, New York, USA.
Macz-Pop, G.A., Rivas-Gonzalo, J.C., Pérez-Alonso, J.J. and González-Paramás, A.M. 2006. Natural occurrence of free anthocyanin aglycones in beans (*Phaseolus vulgaris* L.). Food Chem. 94: 448–456.
Marinova, D. and Ribarova, F. 2007. HPLC determination of carotenoids in Bulgarian berries. J. Food Compos. Anal. 20: 370–374.
Mazza, G., Cacace, J.E. and Kay, C.D. 2004. Methods of analysis for anthocyanins in plants and biological fluids. J. AOAC Int. 87: 129–145.
Mazzuca, P., Ferranti, P., Picariello, G., Chianese, L. and Addeo, F. 2005. Mass spectrometry in the study of anthocyanins and their derivatives: differentiation of *Vitis vinifera* and hybrid grapes by liquid chromatography/electrospray ionization mass spectrometry and tandem mass spectrometry. J. Mass Spectrom. 40: 83–90.
Meléndez-Martínez, A.J., Britton, G., Vicario, I.M. and Heredia, F.J. 2006. HPLC analysis of geometrical isomers of lutein epoxide isolated from dandelion (*Taraxacum officinale* F. Weber ex Wiggers). Phytochemistry. 67: 771–777.
Mendiola, J.A., Martín-Álvarez, P.J., Señoránas, F.J., Reglero, G., Capodicasa, A., Nazzaro, F., Sada, A., Cifuentes, A. and Ibáñez, E. 2010. Design of natural food antioxidant ingredients through a chemometric approach. J. Agr. Food Chem. 58: 787–792.
Mercadante, A.Z., Britton, G. and Rodriguez-Amaya, D.B. 1998. Carotenoids from Yellow Passion Fruit (*Passiflora edulis*). J. Agric. Food Chem. 46: 4102–4106.
Merzlyak, M.N., Gitelson, A.A., Chivkunova, O.B. and Rakitin, V.Y. 1999. Non-destructive optical detection of pigment changes during leaf senescence and fruit ripening. Physiol. Plantarum 106: 135–141.
Merzlyak, M.N., Melo, T.B. and Naqvi, K.R. 2008. Effect of anthocyanins, carotenoids, and flavonols on chlorophyll fluorescence excitation spectra in apple fruit: signature analysis, assessment, modelling, and relevance to photoprotection. J. Exp. Bot. 59: 349–359.
Minguez-Mosquera, M.I., Gandul-Rojas, B. and Fernandez, J.G. 1996. Preparation of Cu(II) complexes of oxidized chlorophylls and their determination by thin-layer and high-performance liquid chromatography. J. Chromatogr. A. 731: 261–271.
Moshou, D., Wahlen, S., Strasser, R., Schenk, A. and Ramon, H. 2003. Apple mealiness detection using fluorescence and self-organising maps. Comput. Electron Agr. 40: 103–114.
Moshou, D., Wahlen, S., Strasser, R., Schenk, A., De Baerdemaeker, J. and Ramon, H. 2005. Chlorophyll fluorescence as a tool for online quality sorting of apples. Biosyst. Eng. 91: 163–172.

Moβhammer, M.R., Stintzing, F.C. and Carle, R. 2005. Colour studies on fruit juice blends from *Opuntia* and *Hylocereus* cacti and betalain-containing model solutions derived there from. Food Res. Int. 38: 975–981.

Müller, H. 1997. Determination of the carotenoid content in selected vegetables and fruit by HPLC and photodiode array detection. Z. Lebensm. Unters. F. A. 204: 88–94.

Nayak, C.A., Chethana, S., Rastogi, N.K. and Raghavarao, K. 2006. Enhanced mass transfer during solid-liquid extraction of gammairradiated red beetroot. Radiat. Phys. Chem. 75: 173–178.

Nicolaï, B.M., Beullens, K., Bobelyn, E., Peirs, A., Saeys, W., Theron, K.I. and Lammertyn, J. 2007. Nondestructive measurement of fruit and vegetable quality by means of NIR spectroscopy: a review. Postharvest Biol. Tech. 46: 99–118.

Nyman, N.A. and Kumpulainen, J.T. 2001. Determination of anthocyanidins in berries and red wine by high-performance liquid chromatography. J. Agric. Food Chem. 49: 4183–4187.

Omata, T. and Murata, N. 1983. Preparation of chlorophyll-a, chlorophyll-b and bacteriochlorophyll-a by column chromatography with DEAE-sepharose CL-6B and sepharose CL-6B. Plant Cell Physiol. 22: 1093–1100.

Ornelas-Paz, J.J., Yahia, E.M. and A.A. Gardea, A.A. 2008. Changes in external and internal color during postharvest ripening of 'Manila' and 'Ataulfo' mango fruit and relationship with carotenoid content determined by liquid chromatography–APcI+-time-of-flight mass spectrometry. Postharvest Biol. Tech. 50: 145–152.

Palozza, P. and Krinsky, N.I. 1991. The inhibition of radical-initiated peroxidation of microsomal lipids by both alpha-tocopherol and beta-carotene. Free Radical Bio. Med. 11: 407–414.

Pavokovic, D. and Krsnik-Rasol, M. 2011. Complex biochemistry and biotechnological production of betalains. Food Technol. Biotech. 49: 145–155.

Peirs, A., Schenk, A. and Nicolaï, B.M. 2005. Effect of natural variability among apples on the accuracy of Vis-NIR calibration models for optimal harvest date predictions. Postharvest Biol. Tech. 35: 1–13.

Peng, Y., Ma, C., Li, Y., Leung, K.S.-Y., Jiang, Z.-H. and Zhao, Z. 2005. Quantification of Zeaxanthin dipalmitate and total carotenoids in Lycium fruits (*Fructus Lycii*). Plant Food Hum.Nutr. 60: 161–164.

Pflanz, M. and Zude, M. 2008. Spectrophotometric analyses of chlorophyll and single carotenoids during fruit development of tomato (*Solanum lycopersicum* L.) by means of iterative multiple linear regression analysis. Appl. Optics 47: 5961–5970.

Piattelli, M. 1981. The betalains: structure, biosynthesis, and chemical taxonomy. pp. 557–573. *In:* Stumpf, W. and Conn, P.M. [eds.]. The Biochemistry of Plants, Vol 7, Secondary Plant Products.Academic Press, New York, USA.

Piatelli, M. and Minale, L. 1964. Pigments of Centrospermae. III. Distribution of betacyanins, Phytochemistry. 3: 547–557.

Plata, N., Konczak-Islam, I., Jayram, S., McClelland, K., Woolford, T. and Franks, P. 2003. Effect of methyl jasmonate and p-coumaric acid on anthocyanin composition in a sweet potato cell suspension culture. Biochem. Eng. J. 14: 171–177.

Pól, J., Hyötyläinen, T., Ranta-Aho, O. and Riekkola, M.-L. 2004. Determination of lycopene in food by on-line SFE coupled to HPLC using a single monolithic column for trapping and separation. J. Chromatogr. A. 1052: 25–31.

Powrie, W.D. and Fennema, O. 1963. Electrophoretic separation of beet pigments. J. Food Sci. 28: 214–216.

Ramin, A.A., Prange, R.K., DeLong, J.M. and Harrison, P.A. 2008. Evaluation of relationship between moisture loss in grapes and chlorophyll fluorescence measured as F_0 (F-α) reading. J. Agric. Sci. Technol. 10: 471–479.

Rastogi, N.K., Eshtiaghi, M.N. and Knorr, D. 1999. Accelerated mass transfer during osmotic dehydration of high intensity electrical field pulse pre-treated carrots. J. Food Sci. 64: 1020–1023.

Rebecca, O.P.S., Boyce, A.N. and Chandran, S. 2010. Pigment identification and antioxidant properties of red dragon fruit (*Hylocereus polyrhizus*). Afr. J. Biotechnol. 9: 1450–1454.
Ren, D. and Zhang, S. 2008. Separation and identification of the yellow carotenoids in *Potamogeton crispus* L. Food Chem. 106: 410–414.
Revilla, E., Ryan, J.M. and Martín-Orteg, G. 1998. Comparison of several procedures used for the extraction of anthocyanins from red grapes. J. Agric. Food Chem. 46: 4592–4597.
Revilla, I., Pérez-Magariño, S., González-SanJosé, M.L. and Beltrán, S. 1999. Identification of anthocyanin derivatives in grape skin extracts and red wines by liquid chromatography with diode array and mass spectrometric detection. J. Chromatogr. A. 847: 83–90.
Romero, G.G., Martinez, C.C., Alanis, E.E., Salazar, G.A., Broglia, V.G. and Alvarez, L. 2009. Bio-speckle activity applied to the assessment of tomato fruit ripening. Biosyst. Eng. 103: 116–119.
Scalzo, R.L., Genna, A., Branca, F., Chedin, M. and Chassaigne, H. 2008. Anthocyanin composition of cauliflower (*Brassica oleracea* L. var. botrytis) and cabbage (*B. oleracea* L. var. capitata) and its stability in relation to thermal treatments. Food Chem. 107: 136–144.
Schiedt, K. 1995. Chromatography: Part III. Thin-layer chromatography. pp. 131–144. *In:* Britton, G., Liaaen-Jensen, S. and Pfander, H. [eds.]. Carotenoids, Vol. 1A, Isolation and Analysis.Birkhauser Verlag, Basel, Switzerland.
Schliemann, W., Kobayashi, N. and Strack, D. 1999. The decisive step in betaxanthin biosynthesis is a spontaneous reaction. Plant Physiol. 119: 1217–1232.
Schmitz, H.H., Emenhiser, C. and Schwartz, S.J.. 1995. HPLC separation of geometric carotene isomers using a calcium hydroxide stationary phase. J. Agr. Food Chem. 43: 1212–1218.
Schoefs, B. 2002. Chlorophyll and carotenoid analysis in food products. Properties of the pigments and methods of analysis. Trends Food Sci. Tech. 13: 361–371.
Schoefs, B. 2004. Determination of pigments in vegetables. J. Chromatogr. A. 1054: 217–226.
Schoefs, B. 2005. Plant pigments: Properties, analysis, degradation. Adv. Food Nutr. Res. 49: 41–91.
Schwartz, S.J. 1988. Pigment analysis. pp. 293–304. *In:* Nielsen, S.S. [ed.]. Food Analysis.2nd ed., Aspen Publishers, Maryland, USA.
Schwarz, M., Hillebrand, S., Habben, S., Degenhardt, A. and Winterhalter, P. 2003. Application of high-speed countercurrent chromatography to the large-scale isolation of anthocyanins. Biochem. Eng. J. 14: 179–189.
Schwartz, S.J. and Lorenzo, T.V. 1990. Chlorophylls in foods. Crit. Rev. Food Sci. 29:1–17.
Schwartz, S.J. and Patroni-Killam, M. 1985. Detection of cis-trans carotene isomers by two-dimensional thin-layer and high-performance liquid-chromatography. J. Agr. Food Chem. 33: 1160–1163.
Scotter, M.J. 2011. Methods for the determination of European Union-permitted added natural colours in foods: a review. Food Addit.Contam. 28: 527–596.
Seeram, N.P. and Nair, M. 2002. Inhibition of lipid peroxidation and structure-activity related studies of the dietary constituents anthocyanins, anthocyanidins, and catechins. J. Agr. Food Chem. 50: 5308–5312.
Sentanin, M.A. and Amaya, D.B.R. 2007. Carotenoid levels in papaya and peach determined by high performance liquid chromatography. Ciencia Tecnol. Alime. 27: 13–19.
Solovchenko, A.E., Chivkunova, O.B., Merzlyak, M.N. and Gudkovsky, V.K. 2005. Relationships between chlorophyll and carotenoid pigments during on- and off-tree ripening of apple fruit as revealed non-destructively with reflectance spectroscopy. Postharvest Biol. Tech. 38: 9–17.
Solovchenko, A.E., Merzlyak, M.N. and Pogosyan, S.I. 2010. Light-induced decrease of reflectance provides an insight in the photoprotective mechanisms of ripening apple fruit. Plant Sci. 178: 281–288.
Sosnova, V. 1970. Reproduction of sugar beet mosaic and tobacco viruses in anthocyanized beet plants. Biol. Plantarum 12: 424–427.

Spanos, G.A., Chen, H. and Schwartz, S.J. 1993. Supercritical CO_2 extraction of ß-carotene from sweet potatoes. J. Food Sci. 58: 817–820.
Stafford, H.A. 1994. Anthocyanins and betalains: evolution of the mutually exclusive pathways. Plant Sci. 101: 91–98.
Stintzing, F.C. and Carle R. 2004. Functional properties of anthocyanins and betalains in plants, food, and in human nutrition. Trends Food Sci. Tech. 15: 19–38.
Stintzing, F.C. and Carle, R. 2007. Betalains—Emerging prospects for food scientists. Trends Food. Sci. Tech. 18: 514–525.
Stintzing, F.C., Conrad, J., Klaiber, I., Beifuss, U. and Carle, R. 2004. Structural investigations on betacyanin pigments by LC NMR and 2D NMR spectroscopy. Phytochemistry 65: 415–422.
Strack, D., Vogt, T. and Schliemann, W. 2003. Recent advances in betalain research. Phytochemistry. 62: 247–269.
Strain, H.H., Cope, B.T. and Svec, W.A. 1971. Analytical procedures for the isolation, identifications, estimation and investigation of the chlorophylls. Methods Enzymol. 23: 452–476.
Streitweiser, A. and Heathcock, C.H. 1981. Introduction to Organic Chemistry. MacMillan, New York. USA.
Stuppner, H. and Egger, R. 1996. Application of capillary zone electrophoresis to the analysis of betalains from *Beta vulgaris*. J. Chromatogr. A. 735: 409–413.
Sugawara, T. and Igarashi, K. 2008. Cultivar variation and anthocyanins and rutin content in sweet cherries (*Prunus avium* L.). J. Jpn. Soc. Food Sci. 55: 239–244.
Suzuki, N., Saitoh, K. and Adachi, K. 1987. Reverse-phase high performance thin-layer chromatography and column liquid chromatography of chlorophylls and their derivatives. J. Chromatogr. 408: 181–190.
Takeoka, G.R., Dao, L.T., Full, G.H., Wong, R.Y., Harden, L.A., Edwards, R.H. and Berrios, J.J.1997. Characterization of black bean (*Phaseolus vulgaris* L.) anthocyanins. J. Agr. Food Chem. 45: 3395–3400.
Taungbodhitham, A.K., Jones, G.P., Wahlqvist, M.L. and Briggs, D.R. 1998. Evaluation of extraction method for the analysis of carotenoids in fruits and vegetables. Food Chem. 63: 577–584.
Tian, Q., Giusti, M.M., Stoner, G.D. and Schwartz, S.J. 2005. Screening for anthocyanins using high-performance liquid chromatography coupled to electrospray ionization tandem mass spectrometry with precursor-ion analysis, product-ion analysis, common-neutral-loss analysis, and selected reaction monitoring. J. Chromatogr. A 1091:72–82.
Timkhum, P. and Terdwongworakul, A. 2012. Non-destructive classification of durian maturity of 'Monthong' cultivar by means of visible spectroscopy of the spine. J. Food Eng. 112: 263–267.
Valverde, J. and This, H. 2008. ^{1}H NMR quantitative determination of photosynthetic pigments from green beans (*Phaseolus vulgaris* L.). J. Agr. Food Chem. 56: 314–320.
Vásquez-Caicedo, A.L., Sruamsiri, P., Carle, R. and Neidhart, S. 2005. Accumulation of all-trans-β-carotene and its 9-cis and 13-cis stereoisomers during postharvest ripening of nine Thai mango cultivars. J. Agr. Food Chem.53: 4827–4835.
Vega, P.J., Balaban, M.O., O'Keefe, S.F. and Cornell, J.A. 1996. Supercritical carbon dioxide extraction efficiency for carotenes from carrots by RSM. J. Food Sci. 61: 757–759.
Venturello, A., Ceccarelli, R., Garone, E. and Geobaldo, F. 2012. Fast non-destructive determination of chlorophylls in apple skin. Italian J. Food Sci. 24: 167–172.
Vogelmann, T.C. 1993. Plant tissue optics. Annu. Rev. Plant Phys. 44: 231–251.
von Elbe, J.H., Sy, S.H., Maing, I.Y. and Gabelman, W.H. 1972. Quantitative analysis of betacyanins in red table beets (*Beta vulgaris*). J. Food Sci. 37: 932–934.
Wang, H., Nair, M.G., Strasburg, G.M., Chang, Y., Booren, A.M., Gray, J.I. and DeWitt, D.L. 1999. Antioxidant and antiinflammatory activities of anthocyanins and their aglycon, cyanidin, from tart cherries. J. Nat. Prod. 62: 294–296.

Wang, S.Y. and Jiao H. 2000. Scavenging capacity of berry crops on superoxide radicals, hydrogen peroxide, hydroxyl radicals, and singlet oxygen. J. Agr. Food Chem. 48: 5677–5684.
Wang, H., Cao, G.H. and Prior, R.L. 1997. Oxygen radical absorbing capacity of anthocyanins. J. Agr. Food Chem. 45: 304–309.
Wesley, J.W.F., Scott, W.T. and Holt, S. 1970. Chlorophyllides-c. Can. J. Biochem. 48: 376–383.
Wiley, R.C. and Lee, Y.N. 1978. Recovery of betalaines from red beets by a diffusion-extraction procedure. J. Food Sci. 43: 1056–1058.
Yang, M., Koo, S.I., Song, W.O. and Chun, O.K. 2011. Food matrix affecting anthocyanin bioavailability: review. Curr. Med. Chem. 18: 291–300.
Xu, Z., Joenathan, C. and Khorana, B.M. 1995. Temporal and spatial properties of the time – varying speckles of botanical specimens. Opt. Eng. 34(5): 1487–1502.
Zanatta, C.F., E. Cuevas, F.O. Bobbio, P. Winterhalter, and A.Z. Mercadante. 2005. Determination of anthocyanins from Camu-camu (*Myrciaria dubia*) by HPLC-PDA, HPLC-MS, and NMR. J. Agr. Food Chem. 53:9531–9535.
Zdunek, A. and Herppich, W.B. 2012. Relation of biospeckle activity with chlorophyll content in apples. Postharvest Biol. Tech. 64: 58–63.
Zhuang, H., Hildebrand, D.F. and Barth, M.M. 1995. Senescence of broccoli buds is related to changes in lipid peroxidation. J. Agr. Food Chem. 43: 2585–2591.
Ziosi, V., Noferini, M., Fiori, G., Tadiello, A., Trainotti, L., Casadoro, G. and Costa, G. 2008. A new index based on Vis spectroscopy to characterize the progression of ripening in peach fruit. Postharvest Biol. Tech. 49: 319–329.
Zude, M. 2003. Comparison of indices and multivariate models to non-destructively predict the fruit chlorophyll by means of visible spectrometry in apple fruit. Anal. Chim. Acta. 481: 119–126.
Zude, M., Birlouez-Aragon, I., Paschold, P.J. and Rutledge, D.N. 2007. Non-invasive spectrophotometric sensing of carrot quality from harvest to consumption. Postharvest Biol. Tech. 45: 30–37.
Zude, M., Pflanz, M., Spinelli, L., Dosche, C. and Torricelli, A. 2011. Non-destructive analysis of anthocyanins in cherries by means of Lambert-Beer and multivariate regression based on spectroscopy and scatter correction using time-resolved analysis. J. Food Eng. 103: 68–75.

6

Lipids in Meat and Seafood

Rui Pedrosa,[1] Carla Tecelão[1,2] and Maria M. Gil[1],*

ABSTRACT

Lipids are a vital nutrient with many functions in all the live cells. Lipids are essential macromolecules of cellular membranes, organelle membranes, one of the main body energy reserves, and are involved in heat regulation. Moreover, lipids are precursors of hormonal synthesis and are involved in several signal transduction pathways, for example, they are associated with the inflammatory process. This would explain why lipid homeostasis dramatically influences human health. Additionally, lipid contents, particularly the composition of saturated fatty acids of the diet are strongly related to several human diseases, such as obesity, cardiovascular diseases, type-2 diabetes, cancer, neurological disorders, etc. Contrarily, polyunsaturated fatty acids, particularly the omega-3 fatty acids, have positive effects against almost all of the pathologies mentioned before.

The fat and fatty acids contents are diverse and very different in meat, meat products and seafood, which can be also critical in food processing and cooking. For consumers, researchers, but also for food processing industries it is critical to know the fat contents and the fatty acid profile of the meat, meat products, fish, fish products and other seafood (e.g., seaweeds) or seafood products. In other words, it is very important to quantify fats in all kind of food or ingredients used in the food industry. The

---

[1] Marine Resources Research Group, School of Tourism and Maritime Technology–Polytechnic Institute of Leiria, Campus 4—Santuário N.ª Sra. dos Remédios, 2520—641 Peniche, Portugal.

[2] CEER-Biosystems Engineering, Instituto Superior de Agronomia–Technical University of Lisbon, Tapada da Ajuda, 1349-017 Lisboa, Portugal.

* Corresponding author

missing or omitted information, for example, the exclusion of the presense of the omega-3 fatty acids in food or food products, is not acceptable to a consumer or from a food industry point of view. On the other hand fatty acid composition in servings is strongly influenced by the quality of fats used during the food processing. Thus, this would agree with the fact that differences in fatty acid profile of oil (fish or vegetable oil) used in the food industry can dramatically change the food process and the nutritional quality of the final food product.

Analysis of lipid extracts from food samples is an important determination in biochemical, physiological and nutritional studies and, therefore, should be carried out with accuracy. However, despite the considerable number of lipid extraction and quantification methods, there is a lack of studies regarding establishment of criteria for choosing the most appropriate one.

This chapter highlights only some of the most representative methods regarding lipid extraction as well as the total fat quantification and characterization of the fatty acid profile and oxidative stability in meat, fish and seaweed samples.

6.1 Introduction

6.1.1 Main Roles and Structure of Lipids

Lipids are one of the key macromolecules of life and play crucial and transversal biological roles in all live organisms: they are the major component of cellular membranes; are important fuel molecules; one of the main means of energy storage; are present in important signal transduction pathways; are precursors for hormonal synthesis, transport and also contribute towards the absorption of fat-soluble vitamins, etc. Apart from the nutritional, physiological and regulatory function, lipids are directly or indirectly (as vehicle) linked to flavors and aromatic substances. There is a general agreement that lipids are critical for food, food processing and in cooking (Lichtenstein et al. 1998; Berg et al. 2007; Schmid 2010).

Lipids are a diverse group of compounds that have as main chemical signature the insolubility in water and are highly soluble in organic solvents such as chloroform, benzene, hexane, etc. These highly hydrophobic proprieties are mainly related to the presence of fatty acids. Fatty acids are carboxylic acids with hydrocarbon chains from 4 to 36 carbons long (C4 to C36) that can be fully saturated or unsaturated with one (monounsaturated fatty acid—MUFA) or several (polyunsaturated fatty acid—PUFA) double bonds. The double bonds in MUFA adopt a typical chemical pattern between C_9 and C_{10}. When we look at the PUFA fatty acids, with the exception of arachidonic fatty acid, the additional PUFA double bonds are located, generally, between C_{12} and C_{13} and C_{15} and C_{16}. Several conventions are used to indicate the position of the unsaturated location. One of the most classical conventions used for the double bond position is established on the fatty acid terminal methyl carbon designation, that is known as ω (omega) or n-carbon. The unsaturated fatty acids are divided into three main series or families: the ω-3 (omega 3) or n-3, ω-6 (omega 6) or n-6 and ω-9 (omega 9) or n-9.

Almost all of the unsaturated fatty acids (UFA) have double bonds with *cis* configuration. The *trans* UFA's are produced in the animals' rumens and the human intake occurs via meat and dairy foods. On the other hand, *trans* UFA's may be generated by hydrogenation in industrial or non-industrial environments by catalytic or non-catalytic process. This apparently minor transformation has dramatic implications for metabolic, physiological and cardiovascular diseases. In fact, diets with high contents of *trans* fatty acids are correlated with increases in blood levels of "bad cholesterol" —low density lipoproteins (LDL) and decreases in blood levels of "good cholesterol"—high density lipoproteins (HDL) (Nelson and Cox 2005; Berg et al. 2007).

6.1.2 Lipids in Meat

As previously mentioned, it is very well-known and clear that lipids play a central role in the human diet. However, lipids can be looked upon as a paradox because they are also the nutrient strongly connected to human diseases. Today, one of the biggest societal health problems, mainly found in industrial countries, is linked not to the transmissible or infectious diseases, but to the non-transmissible diseases like obesity and cardiovascular diseases. Unfortunately, the association of lipids and disease is also true for type-2 diabetes and cancer (Schmid 2010; Ospina et al. 2012). All of these situations are strongly connected to the high fat intakes that are mainly linked to fresh and processed meat. However, this is not absolutely true, because we can find lean meat with high muscle contents that is species, diet, environmental conditions, age and gender-dependent (Pickova 2009; Schmid 2010). Moreover, fat content may be reduced by the cooking or just by cutting the fat off before consumption. The total fat contents found in meat and meat products can reveal very high differences and we can find both low-fat meat and high-fat meat. A very good example is the chicken breast without skin and sausages, which can have lower than 1 g of fat/100 g of meat and up to 40 g of fat/100 g food, respectively. In line with this, different meat parts of the same animal can have different fat contents (e.g., 4 g of fat/100 g of meat-pork fillet; 10 g of fat/100 g of meat-pork chop) (Schmid 2010).

It is well established that the lipid-associated diseases are related mainly to the high saturated fatty acids and cholesterol contents in meat. In fact, the saturated fatty acids (SFA) contents in meat can be more than 50% of the total fatty acids (van Poppel et al. 1998). The SFA mostly present in meat are palmitic acid (16:0; 25–33% of fatty acid content), stearic acid (18:0; 10–20% of fatty acid content) and myristic acid (14:0; 2–3% of fatty acid content) (Enser et al. 1996; Diaz et al. 2005; Valsta et al. 2005). This is consistent with several international authorities' strategies (Food and Agriculture Organization of the United Nations—FAO; World Health Organization—WHO; European Food Safety Authority—EFSA; World Cancer Research Fund—WCRF) that recommend lowering intake of SFA by reducing the meat production and consumption in the industrialized countries. This view is also in agreement with the food industries' strategies to increase efforts to reduce the SFA presence in meat or meat products by improving the fatty acid profile in order to reduce lipid-related diseases, mainly the cardiovascular disease (Ospina et al. 2012). The meat fatty acid profile is also highly marked by the presence of MUFA that typically vary between 40 and 50% of the total fatty acid contents (Diaz et al. 2005). Oleic acid (18:1) is not only the most abundant MUFA, but is also frequently the most abundant component in the meat fatty acid profile (Diaz et al. 2005). On the other hand, PUFA contents in fresh meat can vary between 10 and 30% of the total fatty acid contents (Diaz et al. 2005; Schmid 2010). By contrast, the PUFA contents in meat products is lower that 10% of the total fatty acids by technological reasons (to increase the melting point) and also to guarantee the oxidative stability of the food product (Schmid 2010). This issue is an important bottleneck for food engineering, needing to put into the same box the technological food processes and the nutritional quality related to the meat fatty acid profile.

Two different ratios are very useful and important in the evaluation of the nutritional quality of the meat, the ratio between PUFA:SFA and the ratio of n-6:n-3 fatty acids. The PUFA:SFA ratio values from 1 to 1.5 are considered very good for meat or meat products (Maid-Konhert 2002). However, these values can change dramatically in meat from different animals. The PUFA:SFA ratio can assume very low values in beef (0.1), lamb (0.2), veal (0.2), pork (0.3) and high ratios in ostrich (1.4), rabbit (1.0), horse (1.2) and turkey (1.1) (Enser et al. 1996; Jakobsen 1999; Schmid 2010). On the other hand, it is well known that n-6:n-3 ratios exceeding 5 in human diet are not good from the nutritional point of view, being normally associated with lipid-related diseases. In this matter we can find meat fat with very good n-6:n-3 ratios (e.g., veal and lamb), but also meat fat with n6:n-3 ratios much higher than 5 (e.g., chicken, duck and turkey) (Jakobsen 1999; Schmid 2010; Pestana et al. 2012). Nevertheless, we cannot forget that the n-6:n-3 and PUFA/SFA ratios in meat from the same kind of animal depends on several factors, such as breed, diet, gender, age, etc.

As previously mentioned in the Section 6.1.1, the *trans*-fatty acids are particularly linked to lipid-related diseases. The sources of *trans*-fatty acids are mainly associated with hydrogenation or partial hydrogenation of liquid vegetable oils during food processing. The great interest of the hydrogenation process for food industry is to produce "solid" (at room temperature) oils from liquid oils. In other words, the introduction of *trans*-fatty acids increases the melting point when compared to the *cis*-fatty acids and decreases the melting point when compared to the SFA. The other large source of *trans*-fatty acids are dairy and meat products, mainly associated with ruminants. The *trans*-fatty acids are produced by bacteria present in the rumens of those animals (Remig et al. 2010). *Trans*-fats from natural or commercial (chemical hydrogenation) sources are vaccenic acid (11-*trans* 18:1) and conjugated linoleic acid (9-*cis*, 11-*trans* 18:2; CLA) or elaidic acid (9-*trans*18:1) that assume different structures (Remig et al. 2010). However, several studies showed that *trans*-fatty acids from either a natural source or a commercial source had no adverse effects on HDL, LDL and cholesterol levels (Chardigny et al. 2008; Motard-Belanger et al. 2008). However, the intake of natural *trans*-fatty acids is very low when compared to the amounts of consumed *trans*-fats that are generated by hydrogenation. In fact, several evidences support the view that CLA (natural *trans* fatty acid) present in lean red meat has anti-carcinogenic, anti-inflammatory and anti-atherosclerotic effects (Ferguson 2010; McAfee et al. 2010). On the other hand, several studies associate high intake of meat, especially red meat and processed meat, with an increased risk of cancers, especially colorectal cancer. However, some researchers assume that this risk may not be a function of meat *per se*, but may reflect high fat intake and/or carcinogens generated through the cooking and processing methods (Ferguson 2010). We cannot forget that red meat is an important source of selenium, vitamin B6, vitamin B12 and vitamin D. Therefore, finding the balance between meat and other foods may be crucial against some of the cancer risks.

6.1.3 Lipids in Seafood

The main lipid signatures found in fish are related to the presence of the long-chain PUFA and can be considered antagonic lipid fingerprints when compared to the lipid signature present in meat. Fish are the most important dietary source of highly PUFA, such as the omega-3 docosahexaenoic acid (DHA) and eicosapentaenoic acid (EPA). However, fish are incapable of synthetizing these long chains of PUFA that are diet-dependent through microalgae, macro-algae and small fish present in their diet (Lunn and Theobald 2006). There is a universal agreement that the fish lipid contents and particularly the amounts of the long-chain omega-3 PUFA (e.g., DHA and EPA) may vary strongly depending on the diet (including food availability), on the fish species, geographical location, season, water temperature, age, size and maturation status (Schwalme et al. 1993; Bandarra et al. 1997; Ould Ahmed Louly et al. 2011; Tufan et al. 2011; Strobel et al. 2012). Moreover, the lipid distribution is also very different throughout the fish body. In cod, for example, the main fat reserves are maintained in the liver and not in the muscle fibrils, as in the fatty fish (Tocher et al. 2006; Strobel et al. 2012).

According to Ackman (1990) fish species may be grouped into four categories: lean fish (<2% of total fat), low-fat (2–4% of total fat), medium-fat (4–8% of total fat) and high-fat or oily fish (>8%). Nowadays, this nutritional fish classification (according to the fat contents) is still used.

The total fat contents in fish may vary from 0.2% (garfish) to 40% (cod liver). Still, for the same edible part of same species, the fat content may range from 0.4 to 18.4 (observed for sardines, the *Sardine plichardus*) or from 1.8 to 23 (observed for mackerel—*Scomber scombrus*) according to the season and maturation status (Bandarra et al. 1997; Soriguer et al. 1997). Moreover, for the same species the fat contents may change dramatically, not only between organs such as muscle or liver, but also depending on the food processing method used. A great example is that of cod fillets, smoked cod and cod liver that may have 0.56%, 3.39% and 40.4% of fat, respectively (Strobel et al. 2012).

The total SFA can range from 20.5% to 53% of the total fatty acid profile (Huang et al. 2012; Prato and Biandolino 2012). The dominant SFA in fish is palmitic acid (C16:0), which may contribute up to 70% of the total SFA (Zlatanos and Laskaridis 2007; Ozogul et al. 2009; Prato and Biandolino 2012). The second dominant SFA is stearic acid (C18:0) that balances from 1% to 11.6% (Huynh and Kitts 2009; Prato and Biandolino 2012; Strobel et al. 2012).

Generally, the oleic acid is the most abundant MUFA in most marine fish species, accounting for 60–75% of the total MUFA (Ozogul et al. 2009; Prato and Biandolino 2012). The MUFA fatty acids may vary from 17.8% (sand smelt—*Atherina boyeri*) to 53% (herring—*Clupea harengus*) of the total fatty acid profile (Prato and Biandolino 2012; Strobel et al. 2012).

Essentially, as previously mentioned, the main fatty acid fingerprint of fish species is normally described by high amounts of PUFA, particularly associated to the long-chain omega-3 PUFA, such as EPA and DHA. Actually, several studies showed that PUFA accounts for more than 30% (24 of 38 species studied) of the total fatty acids (Huynh and Kitts 2009; Huang et al. 2012; Prato and Biandolino 2012; Strobel et al. 2012). However, for some of the species studied, the PUFA fatty acids represented more than 50% of the total fatty acids (*Mugil cephalus, Scieaena umbra, Gadus merlangus,*

Pollachius virens and *Sardinops sagax*). Moreover, these same studies also demonstrated that the omega-3 fatty acids represented more than 20% of the total fatty acids (33 of 38 different species studied). In agreement with this, PUFA contents are largely dominated by the presence of DHA, in some species it can account for up to 40% of the PUFA fatty acids. Generally EPA is the second typical PUFA that may balance between 5 and 12% of the total PUFA (Huynh and Kitts 2009; Huang et al. 2012; Prato and Biandolino 2012; Strobel et al. 2012).

As mentioned before, low n-6:n-3 (< 5) and high PUFA:SFA (> 1) ratios are associated with high nutritional value of foods. The n-6:n-3 ratio in fish varies significantly among fish species. However, fish species normally present a very favorable n-6:n-3, which is normally lower than one. According to this, fish also has a very good PUFA:SFA ratio value, higher than one (Sirot et al. 2008; Huynh and Kitts 2009; Huang et al. 2012; Strobel et al. 2012).

Naturally, the species with the highest amount of DHA and EPA n-3 PUFA are the high fat or oily fish. However, the highest percentage of PUFAs DHA and EPA are normally present in lean fish (Prato and Biandolino 2012; Strobel et al. 2012).

The total fat and fatty acid composition may also differ strongly between wild and farmed fish species, mainly due to varying the amount of vegetable and fish oil in fish diet. One of the main results of a nutritional imbalance of the feed is a decrease of the nutritional value of aquaculture fish. This was very well reported by Strobel et al. (2012) who showed the marked differences in wild and farmed salmon. The fat content of aquaculture salmon and wild salmon was 12.3% and 2.07%, respectively. However, wild salmon showed higher percentage of long chain omega-3 PUFA (28.9%) when compared to farmed salmon (21.2%) (Strobel et al. 2012).

Mollusks have the same quantity of fat as lean fish (< 2%). By contrast, crustaceans have much more fat than mollusks (4–7%). Crustaceans are also richer than mollusks in omega-3 PUFA, particularly on DHA (Sirot et al. 2008).

The algae lipid metabolism is unique and has an enormous variety of lipid classes that are not present or synthetized by other live organisms (for details please see Harwood and Gushina 2009). Consequently, it is not surprising that algae have the main role in primary production at the base of the food chain.

The food science interest in algae is strongly associated with edible seaweeds (macro-algae). From a nutritional point of view, seaweeds are good sources of fibers, vitamins (A, B1, B2, B3, B6, B9, B12 and C), minerals and trace elements like iodine (MacArtain et al. 2007). Notwithstanding, the seaweed nutritional hot-spots are clearly linked to the variety, quality and amounts of the PUFA. The total lipid contents of seaweed may vary strongly between species, ranging from 1 to 5 g/100 g (seaweed dry weight). Van Ginneken et al. (2011) found fat contents for *Ascophyllum nodosum* (egg wrack), *Ulva lactuca* (sea lettuce) and *Chondrus crispus* (irish moss or carrageen) of 4.5, 2.2 and 1.0% (seaweed dry weight), respectively.

As previously mentioned, several organizations like FAO, WHO and EFSA recommend a reduction intake of SFA and an increase on the intake of PUFA. In general the PUFA are the class of fatty acids more abundant in seaweeds and normally account for one half of the total fat (MacArtain et al. 2007; van Ginneken et al. 2011). On the other hand, as discussed before, the n-6:n-3 ratio in human diet higher than 5 is undesirable from the nutritional point of view and is normally associated to lipid-related diseases. This suggestion is in line with several organizations that recommended n-6:n-3 ratios less than 5 in order to prevent inflammatory, cardiovascular and nervous system disorders. The seaweeds' n-6:n-3 ratios may vary between 0.05 (*Palmaria palmata*), 0.11 (*Caulerpa taxifolis*) and 0.36 (*Chondrus crispus*) (van Ginneken et al. 2011). This fatty acid profile is even more relevant if EPA contents are considered. Seaweeds like *Undaria pinnatifida*, *Laminaria hyperborea* and *Palmaria palmata* present 16, 26 or 59 % of EPA of the total fatty acid content, respectively (van Ginneken et al. 2011). In line with this, it is not surprising that seaweeds may become a "new" (they have an old tradition and history in the Asian gastronomy) and important food source for industrialized western countries.

One of the main problems related with foods rich in PUFA are the high sensitivity to oxidation. The ability of PUFA, especially those with more than two double bonds, to rapidly oxidize, is a key point for the reduction of the shelf life of food products by rancidification and color deterioration (Kouba and Mourot 2011).

PUFA oxidation may happen through photo-oxidation, autoxidation or by enzymatic mediated oxidation. The lipid peroxidation reactions are triggered by both free radicals (hydroxyl radical; peroxyl radicals) and non-free radical oxidants such as singlet oxygen (Sun et al. 2011). The lipid peroxidation process has three stages: initiation, propagation and termination. The initiation is normally induced by free radicals and may be accelerated by metal ions, heat, light and lipolytic enzymes. Hydroperoxide, the primary product of this oxidative chain reaction, is very unstable and may react with oxygen to form secondary products (including volatile compounds) that accelerate the oxidation of food-coloring matter, flavor substances and vitamins. Moreover, hydroperoxide can generate peroxyl radicals that are the key molecules involved in the propagation of the lipid peroxidation chain reaction (Headlam and Davies 2003; Niki and Yoshida

2005; Niki et al. 2005). The termination of lipid peroxidation is associated with the condensation reaction between hydroperoxides producing dimers and polymers that may also break down and produce secondary volatile and non-volatile compounds that also cause color changes and rancid odor (Sun et al. 2011). There is an agreement that lipid oxidation is a huge problem for the food industry because it is associated with the development of undesirable flavors and toxic products, limiting the shelf life of foods. Moreover, the lipid oxidation in food not only reduces the shelf life, but also changes the texture, taste, appearance and can be an enormous bottleneck for the introduction of PUFA into functional foods (Campo et al. 2006; Sun et al. 2011). A very good strategy to avoid the ability of unsaturated fatty acids, especially those with more than two double bonds, to rapidly oxidize, is the use of hydrophobic antioxidant, such as vitamin E or Co-enzyme Q10 (Co-Q10) in the diet or as an additive in the production of foods or functional foods (Pravst et al. 2010; Kouba and Mourot 2011). These antioxidants break the lipid peroxidation chain reaction by reacting with peroxyl radicals to form non-radical products. Co-Q10 occurs naturally in all the cells that have mitochondria and is not surprising that meat and fish are among the main sources of Co-Q10. One of the strategies to enrich the foods or functional foods with high levels of PUFA with Co-Q10 is the addition of Co-Q10 to food during processing or/and the addition of Co-Q10 into the animal feed (Pravst et al. 2010).

6.1.4 Omega 3 and Health

The n-3 fatty acids are essential compounds involved in several important physiological processes. It is established that omega-3 fatty acids protect against cardiovascular morbidity and mortality, mainly due to hypotriglyceridemic effects (Calder 2009; Siriwardhana et al. 2012). This health benefit is particularly linked to marine-based fish, seaweed and fish oil food products and are mainly associated with EPA and DHA. These two specific n-3 PUFAs have several benefits against cardiovascular diseases (CVDs). Several researchers and organizations support the idea of increasing consumption of fish or fish products in order to improve the intake of EPA and DHA, particularly in the Western and industrialized countries (Calder 2009; Mozaffarian and Wu 2012; Siriwardhana et al. 2012). Various professional bodies and government organizations recommend 500 mg of n-3 PUFA per day (Meyer 2011). The fact that cardiovascular disease rates are much lower in countries like Japan compared with the Western countries that have n-3 PUFA intakes up to 5 fold lower than Japanese intakes (Meyer 2011) supports the above recommendation. This is more critical because the alpha-linolenic acid (omega-3 fatty acid), which is present in vegetables, may only partially be converted into EPA in humans. In contrast, the conversion to DHA appears to be very poor or non-existent in humans. Some of the alpha-linolenic acid-induced health benefits appear to be associated with the EPA synthesis. However, this conversion is very limited and the outcomes are not comparable to the increased intake of preformed EPA + DHA (Calder and Yaqoob 2009b). Some researchers assume that the long omega-3 fatty acids are not synthesized in the human body and must be obtained through diet or by taking dietary supplements (Cassileth 2010).

The health advantages of the n-3 fatty acid intake are also linked to inflammatory disease and arthritis (Calder and Yaqoob 2009a). The omega-6 (n-6) and omega-3 (n-3) PUFAs are precursors of potent lipid mediators, eicosanoids, which play an important role in the regulation of inflammation. Eicosanoids derived from n-6 PUFAs (e.g., arachidonic acid) have pro-inflammatory and immune-active functions, whereas eicosanoids derived from n-3 PUFAs, such as EPA and DHA, have anti-inflammatory properties, traditionally attributed to their ability to inhibit the formation of n-6 PUFA-derived eicosanoids. This is the pathway more strongly associated with the EPA and DHA anti-inflammatory properties. This can be a particular problem in the Western and industrialized countries that have diets with high ratios of n-6:n3 PUFA (Wall et al. 2010). Moreover, the n-3 fatty acids are also associated with the improvements in the childhood learning and behavior and prevent neurodegenerative diseases. DHA, that is not synthetized by humans, has an important structural role in the eye and brain and its supply early in life is known to be of vital importance (Calder 2009; Wall et al. 2010). Several animal studies have shown that long-chain PUFAs like DHA and arachidonic acid have important functional and modulatory roles on the central nervous system (CNS). These PUFAs are present in high concentration in breast milk and several reports associated DHA to the enhanced intellectual development in the breast-fed children. In line with this view, deficit on DHA was associated to neurodevelopmental disorders (Belkind-Gerson et al. 2008). A particular association of DHA with a neurodegenerative disorder is related to the Alzheimer disease. Actually, some epidemiological studies have shown strong connections between the low fish intake and low levels of DHA to an increased risk of Alzheimer disease. Animals submitted to a diet with low levels of DHA showed learning and memory deficits, whereas their brains show inflammatory and oxidative damage to neurons and synaptic defects. These effects were blunted by DHA supplements (Pauwels et al. 2009). DHA neuroprotection mediated effects have been explained by the DHA anti-inflammatory action. In fact, both n-6 and n-3 (mainly DHA) regulate the inflammation process by the eicosanoid synthesis. However, eicosanoids produced from n-6 PUFAs have strong pro-inflammatory action, whereas eicosanoids derived from n-3

PUFAs, such as DHA, have anti-inflammatory properties. In other words, DHA antagonizes pro-inflammatory eicosanoids (e.g., prostaglandin E2) produced by n-6 PUFA (Pauwels et al. 2009; Wall et al. 2010; Siriwardhana et al. 2012). Another anti-inflammatory DHA and EPA-dependent mechanism is associated to the nuclear factor-kB (NF-kB). NF-kB is a transcription factor involved in several signal transduction pathways, some of them related to the pro-inflammatory cytokine production, including interleukin 6 and tumor necrosis factor-α (TNF-α). EPA and DHA decrease the interleukin 6 and TNF-α by decrease in the NF-kB activity (Siriwardhana et al. 2012).

6.2 Lipid Extraction Methods

The extraction of lipids is an important determination in biochemical, physiological and nutritional studies of different types of foods and, therefore, should be carried out with accuracy. It is a critical step in the analysis of total lipids, especially of the composition of fatty acids.

As previously referred, there are several types of lipids depending on their chemical composition. In relation to their extractability, they can be divided into two groups: hydrophobic lipids such as triacylglycerols and cholesterol, usually present as large globules that can be easily extracted by most solvents; and complex lipids, usually constituents of membranes, where they occur in a close association with such compounds as proteins and polysaccharides, with which they interact, which are not extracted so readily. During lipid extraction, there must be disruption of interactions such as van der Waals' forces and covalent bonds, which act to form complexes between lipids and proteins or carbohydrates (Manirakiza et al. 2001). Thus, procedures for the extraction of lipids from fish and meat products must remove them from their binding sites with cell membranes, lipoproteins and glycolipids. Moreover, the solvents (or various combinations of solvents) used for lipid extraction should have a high solubility for all lipid compounds and be sufficiently polar, to completely extract all of the lipid compounds from a sample, while leaving all of the other compounds behind (Christie 1993). Solvents can be broadly classified into two categories: polar and non-polar. Generally, the dielectric constant of a solvent provides a rough measure of its polarity. Solvents with a low dielectric constant are generally considered to be nonpolar, such as chloroform, and are used to break hydrophobic interactions (lipids with hydrophobic chains-triacylglycerols). On the other hand, solvents with a high dielectric constant are generally considered to be polar, such as methanol, and are used to break hydrogen bonds (phospholipids and glycolipids) (Aued-Pimentel et al. 2010).

As previously referred, lipids are insoluble in water, which makes possible the separation of the globular proteins, carbohydrates and water in tissues. The fact that lipids have a wide range of relative hydrophobicity means that it is practically impossible to use a single universal solvent to extract them all. Thus, several solvent systems might be considered, depending on the type of sample and its components. One should be aware that if a determined solvent is chosen, the total lipid content may be different if another solvent is used instead (Smedes and Askland 1999).

A large number of different extraction procedures have been developed for total lipid extraction and may be found in books and articles. They vary in the nature of the solvents, the method of homogenization, removal of contaminants and many other aspects.

In 1879 Franz von Soxhlet described the first method based on an automatic solvent extraction (diethyl ether) for milk lipids (Soxhlet 1879). This methodology is undoubtedly the oldest and the most widespread method for lipid extraction, being one of the procedures adopted as a reference in terms of extraction efficiency.

The next crucial step in the field of amphipathic lipids extraction from animal tissues was made by Folch in 1957. His work shows that lipids may be extracted using a chloroform/methanol/water phase system (Folch et al. 1957). This procedure, and its modifications, are still the classical and most reliable means for quantitative lipid extraction from animal tissue around the world (Iverson et al. 2001). Some other procedures were developed by Bligh and Dyer (1959) and Sheppard (1963), which also used solvent mixtures made of chloroform/methanol and ethanol/diethyl ether, respectively. The Bligh and Dyer method (BD) may be used for dry food or products with high moisture content (like fish and green vegetables, for example). Due to the use of polar solvents, all lipid classes are extracted.

Despite some drawbacks, the BD and Folch method (FOL) are widely used. The toxicity of solvents used and the undesirable contaminants in non-lipid extraction from organic phase can be pointed out as the major disadvantages of those methods.

Lipid content in animal tissues can also be estimated using Soxhlet apparatus which is the official recommended method (AOAC 2005). Furthermore, new Soxhlet extraction systems offer different modes of extraction which could improve the extraction procedure, reducing the extraction time and the solvent volume.

Despite the considerable number of lipid extraction methods, there is a lack of studies establishing criteria for choosing the most appropriate one. The choice of the most appropriate method for extracting lipids depends on the nature of the tissue matrix (e.g., if the sample is of animal, vegetable or microorganism origin), since it can influence the efficiency of the procedure (Christie 1993). Besides this, moisture level, use of additives and processing technologies should also be considered, as they can also compromise the efficiency of the method (Aued-Pimentel et al. 2010). Baily et al. (1994) concluded that the use of different extraction methods results in different lipid recoveries from biological samples.

Lipid extraction methods that have been used for meat and fish are gathered in Table 6.1. Emphasis is given to traditional extraction procedures using organic solvents, however, non-destructive instrumental methods are also presented.

As already mentioned, conventional methods for lipid extraction (Folch (1957); Bligh and Dyer (1959); Hara and Radin procedure (Hara and Radin 1978); modified Bligh and Dyer (Smedes 1999); Jensen method (Jensen et al. 2003)) are widely used. Nevertheless, they have not been much improved, despite modifications in solvent mixtures and laboratory practices. Improvements are still required for reducing the long preparation times and re-extraction steps needed to ensure complete lipid isolation. Besides, waste disposal of solvents is an additional problem that adds extra cost to the analytical procedure, extra damage to the environment and also creates health hazards for the laboratory personnel. Due to environmental concerns and potential health hazards of organic solvents, new technologies such as supercritical fluid extraction (SFE), pressurised liquid extraction (PLE), near infrared spectroscopy (NIR) and microwave assisted extraction (MAE), have been reported for lipid extraction (Table 6.1).

The most suitable methods used to extract lipids from fish and meat will be briefly described in Sections 6.2.2 to 6.2.4.

According to Christie (1993): "…there are three main aspects that should be considered to any practical procedure for extracting lipids from tissue; firstly, exhaustive extraction and solubility of the lipids in organic solvent and secondly, removal of non-lipid contaminants from the extracts; thirdly, the potential toxicity of solvents to analysts." Besides this, sample preparation and pre-treatment (drying, reducing particle size and possibly acid hydrolysis) in order to release the lipid, are of major importance for an effective analysis. Integrity of lipids can be compromised if inappropriate methods are used to store, prepare and pre-treat the samples.

6.2.1 Sample Preparation

The results of chemical analyses can only be as good as the sampling and sample preparation. The adequacy and condition of the sample or specimen received for examination is of primary importance. If samples are improperly collected and insufficiently fresh, lipid oxidation or hydrolysis prior to the lipid extraction may occur, and thus the laboratory results will be meaningless. Moreover, it is of great importance to ensure that no significant changes are observed in its properties from the moment of

Lipids in Meat and Seafood 155

Table 6.1 Overview of lipid extraction methods (adapted from Schlechtriem et al. 2012).

Method	Principle of method	Advantage	Disadvantage	Studies–Fish Samples	Studies–Meat Samples
Folch et al. (FOL) (1957)	Gravimetric quantification using 1-step solvent extraction with mixture of chloroform, methanol and saline solution (8:4:3) followed by a wash with 0.9% potassium chloride.	• Standard method; • Well established to determine total lipids.	• Adverse effects of chloroform on the environment (EU regulation controlling chlorinated solvents); • Laborious (filtration, etc.).	Iverson et al. 2001 Kondo et al. 2005 Rinchard et al. 2007 Nanton et al. 2007 (Modified Folch-Method including butylated hydroxyltoluene as antioxidant) Ramalhosa et al. 2012	Prevolnik et al. 2005 Tanamati et al. 2005 Perez-Palacios et al. 2008 Brum et al. 2009 Perez-Palacios et al. 2012
Bligh and Dyer (BD) (1959)	Gravimetric quantification using 3-step solvent extraction: (1) methanol + chloroform (2) chloroform and (3) water are added to the tissue. After phase separation total lipids are determined in the chloroform phase by gravimetric analysis following evaporation of the solvent. *Recommended by US_EPA for field studies.*	• Standard method; • Determines total lipids; • Samples may be analysed directly with no pre-drying necessary; • Reduction in the solvent/sample ratio; • Recovery of 95% of total lipids.	• Adverse effects of chloroform on the environment (EU regulation controlling chlorinated solvents); • Laborious (filtration, etc.); • High variability when interlaboratory comparisons.	Iverson et al. 2001 Jensen et al. 2003 Ozogul et al. 2012 Ramalhosa et al. 2012 Xiao et al. 2012	Berg et al. 1997 Tanamati et al. 2005 Perez-Palacios et al. 2008 Brum et al. 2009

Table 6.1 contd....

156 *Methods in Food Analysis*

Table 6.1 contd.

Method	Principle of method	Advantage	Disadvantage	Studies–Fish Samples	Studies–Meat Samples
Hara and Radin (HR) (1978)	Gravimetric quantification using 1-step solvent extraction with hexane/isopropanol (3:2) followed by a wash with aqueous sodium sulphate. *Recommended by US_EPA for field studies.*	■ Solvents less toxic and cheaper than chloroform and methanol; ■ No interference in processing by lipoproteins; ■ Extract contains less nonlipids compared to chloroform-methanol extracts of Folch.	■ Laborious; ■ No extraction of gangliosides, a minor fraction of total lipids.	Gunnlaugsdottir and Ackman 1993 Ramalhosa et al. 2012	Tanamati et al. 2005
Smedes method (BDS) (Smedes 1999)	This method is a modification of the Bligh and Dyer procedure in which methanol is replaced by propan-2-ol and chloroform by cyclohexane, respectively. *Standard procedure for QUASIMEME Laboratory Performance Studies*	■ One key advantage of cyclohexane over chloroform was its lower density, consequently it is separated on top of the extraction mixture; ■ Robust enough for routine use; ■ No step-change in international monitoring data which have so far used Bligh and Dyer as standard method; ■ No filtration required; ■ No chlorinated solvents required; ■ Relatively non-toxic solvents.	■ Laborious; ■ The extraction of specific tissues, like liver, may lead to the formation of an emulsion which can be prevented by replacing the water by 1 M perchloric acid to denature the proteins. The addition of sodium chloride may also help.	Manirakiza et al. 2001 Schlechtriem et al. 2003 Jensen et al. 2009 Karl et al. 2012	Tanamati et al. 2005

Method	Description	Advantages	Disadvantages	References
Jensen method (Jensen et al. 2003)	Gravimetric quantification using 3-step solvent extraction: - 2-propanol and diethyl ether - n-hexane/ diethyl ether and 2-propanol - n-hexane/ diethyl ether	• No halogenated solvents; • Gentle method without heating; • Easy to handle; • Gives BD comparable results for fat and lean fish.	• Laborious; • Special glass apparatus required; • Suitability for small samples have to be checked; • No interlaboratory study; • Further validation/ calibration approaches required NB.: The original "Jensen method" (Jensen et al. 1972) uses a different solvent system leading to an underestimation of the lipid content of very lean fish.	Isaac et al. 2005 Jensen et al. 2009	NA
Soxhlet method (SE) (AOAC 1995; 2005)	Gravimetric quantification using solid-liquid extraction in a Soxhlet Apparatus. Constant flow of organic solvent over material. Solvent is boiled, condenses and passes the tissues several times thereby extracting the lipids. After a suitable time the process is stopped, solvent evaporated and fat weighted.	• Simple; • Not very labour intensive; • Can be operated with non-chlorinated solvents; • Lipids can be used for further determinations.	• Results lower than those of Bligh and Dyer method; • Extractable lipids are determined, not total lipids; • Large amounts of solvents needed; • Special equipment required;	Schlechtriem et al. 2003 Ozogul et al. 2012 Ramalhosa et al. 2012 Xiao et al. 2012	Prevolnik et al. 2005 Tanamati et al. 2005 Perez-Palacios et al. 2008 Brum et al. 2009

Table 6.1 contd....

158   *Methods in Food Analysis*

Table 6.1 contd.

Method	Principle of method	Advantage	Disadvantage	Studies–Fish Samples	Studies–Meat Samples
			▪ Possibly adverse effects on labile lipids and test substance by high temperatures and oxygen; ▪ Results are very much operationally dependent (solvent composition, extraction time, cycles); ▪ Conditions are difficult to control (continuous flow of solvents); ▪ Time consuming.		
Supercritical Fluid Extraction (SFE) (King 2002)	Gravimetric quantification using supercritical fluid extraction. Sample is extracted with liquid carbon dioxide* which serves as solvent. After extraction it is allowed to evaporate and the remaining lipids are weighed. *under normal pressure, carbon dioxide is either gaseous or solid. Under pressure it is taken past its critical point and only the fluid state can exist.	▪ Rapid; ▪ No organic solvent or acid needed; ▪ Lipids can be used for further analysis.	▪ Very expensive equipment; ▪ Complex equipment; ▪ Supply of carbon dioxide needed.	Zhou et al. 2010 Rubio-Rodriguez et al. 2012 Sarker et al. 2012	Berg et al. 1997 Vinci et al. 1999 Berg et al. 2002

Near infrared spectroscopy (NIR)	Measurement depends on the absorption of infrared energy at specific wavelength by functional groups such as the carbonyl group in the ester linkage of lipids.	▪ Non-destructive method; ▪ Sample can be used for other measurements e.g., contaminant analysis; ▪ Can be used on whole fish without removing skin and scales; ▪ Accuracy. The values obtained by NIR agree well with those obtained by solvent extraction.	▪ Requires sophisticated equipment; ▪ Not widely available; ▪ Not widely used as a method of determining lipid content.	Mathias et al. 1987 Darwish et al. 1989	Cronin and McKenzie 1990
Werner-Schmid method or Schmid-Bondzynski-Ratzlaff method	Gravimetric quantification using acid hydrolysis followed by solvent extraction. Sample is heated in a water bath with hydrochloric acid. After cooling, lipid is extracted 3-4 times with diethyl ether and petroleum ether. Solvent is evaporated and lipid weighted.	▪ Extracts all lipids; ▪ Cheap; ▪ Samples can be analysed without pre drying.	▪ Triglycerides can be degraded by acid hydrolysis, therefore lipids cannot be used for other determinations (e.g., fatty acid profiles); ▪ Relatively labour intensive; ▪ Large amounts of solvents and special equipment required;	James 1995 McLean and Drake 2002	James 1995 Berg et al. 2002 McLean and Drake 2002 Tanamati et al. 2005

Table 6.1 contd....

160 Methods in Food Analysis

Table 6.1. contd.

Method	Principle of method	Advantage	Disadvantage	Studies–Fish Samples	Studies–Meat Samples
Weibull-Stoldt or Weibull-Berntrop method	Gravimetric quantification using acid hydrolysis followed by Soxhlet extraction. Sample is mixed with hydrochloric acid and boiled for 30 min. Extract is cooled, filtered and filter washed until free of acid. Residue is dried and Soxhlet extracted.	▪ Extract all lipids; ▪ Cheap; ▪ Samples can be analysed without pre drying.	▪ Triglycerides can be degraded by acid hydrolysis, therefore lipids cannot be used for other determinations (e.g., fatty acid profiles); ▪ Relatively labour intensive; ▪ Large amounts of solvents needed; ▪ Special equipment required; ▪ Results are very much operationally dependent (solvent composition, extraction time, cycles)conditions are difficult to control; ▪ Time consuming.	McLean and Drake 2002	BSI - BS4401-4 Arneth 1998 Wu and Kohler 2000 McLean and Drake 2002
Microwave-Assisted Extraction (MAE)	Performed with an open-vessel extraction apparatus similar to the system of Soxhlet. The solvent is an equivolume mixture of ethyl acetate and cyclohexane. The solvent forms a ternary azeotrope with water, and water is separated after re-condensation in a water trap.	▪ Samples can be analysed without pre drying; ▪ Requires less laboratory material, solvent volume and time. ▪ Low toxicity solvents are applied; ▪ Easy to perform with several samples in parallel.	▪ Studies dealing with MAE of lipids are scarce; ▪ Not widely used as a method of determining lipid content.	Batista et al. 2001 Ramalhosa et al. 2012	NA

sampling to the time when the analysis is carried out. In fact, in order to prevent changes in sample matrix properties, extraction of lipids should be performed as soon as possible after the removal of tissues from a living being. If immediate extraction is not possible, samples should be stored at very low temperatures in sealed containers, under an inert (nitrogen) atmosphere or on dry ice (Shahidi and Wanasundara 2008).

6.2.2 Liquid-Liquid Extractions

Folch Method (FOL)

One of the most widespread solvent mixtures used as lipid extractants is the combination of chloroform and methanol. The combination of these solvents allows the extraction of both amphipathic and nonpolar lipids (Shahidi and Wanasundara 2008).

Within the procedure of Folch, it is essential that the ratio of chloroform, methanol and saline solution in the final mixture be close to 8:4:3 otherwise selective losses of lipids may occur.

Procedure (according to Folch et al. 1957)

Homogenize the tissue with chloroform: methanol (2:1) to a final dilution 20 times the volume of the tissue sample, i.e., the homogenate from 1 g of tissue is diluted to a volume of 20 ml. The time of homogenization will vary with the sample but a minimum of 3 minutes is usually required. Filter the homogenate through a suitable paper into a glass-stoppered bottle (centrifugation may be used instead of filtration). For the purpose of computation, this extract corresponds to 0.05 times its volume of tissue, i.e., 1 ml of extract corresponds to 0.05 g of tissue. Wash the crude extract with 0.2 of its volume of either water or salt solution. Allow the solution to separate into two phases. The volumes of the upper and lower phases are 40 and 60% of the total volume respectively. Remove the upper layer by siphoning. Rinse the interface three times with pure 'upper phase', i.e., the chloroform: methanol: water 3:48:47 so that the lower phase is not disturbed. This has the effect of removing any 'fluff' at the interface. Finally add methanol so that the lower phase and the rinsing liquid form one phase. Dilute the resulting solution to any desired volume by the addition of chloroform: methanol (2:1). The last two steps may be omitted if it is intended to remove the solvent under vacuum to yield a dry extract for weighing.

The tissue may be homogenized initially in the presence of both solvents, although better results are often obtained if the methanol (10 ml/g tissue) is added first, followed after brief blending by chloroform (20 volumes).

More than one extraction may be needed, but with most tissues the lipids are removed almost completely after two or three treatments.

The Bligh and Dyer Method (BD)

The Bligh and Dyer method (and its modifications) is one of the most studied methods and it is an adaptation of the FOL method. It was developed as an economical procedure for extracting lipids from samples that have large amounts of water, e.g., frozen fish, with low volume of solvent.

Procedure (adapted from Honeycutt 1995)

Weigh 10 g of wet sample to a pre-weighed 100 ml conical flask. Add 20 ml methanol and 10 ml chloroform and homogenize the sample for 2 min with an UltraTurrax mixer. Add 10 ml of chloroform a second time and homogenize the mixture vigorously for 1 min. Add 10 ml of distilled water and homogenize again for 30 s. Separate the two layers by centrifugation for 10 min at 450 × g in a thermostatic centrifuge at 20 °C. Transfer the lower layer to a pear-shaped flask with a Pasteur pipette. Perform a second extraction with 20 ml 10% (v/v) methanol in chloroform by vortexing for 2 min. After centrifugation, add the lower chloroform phase to the first extract. Evaporate the sample to dryness (e.g., with a rotavapor). The residue will be further dried at 104 °C for 1 h. Record the extracted weight and calculate the lipid content.

One of the most common modifications is to replace water by 1M sodium chloride. This addition blocks the binding of some acidic lipids to denatured lipids.

The Hara and Radin Method (HR)

Considering the potential risks, both to health and to the environment, of the use of chloroform-methanol mixture, Hara and Radin (1978) proposed a simple method, based on the use of a mixture of n-hexane–isopropanol, which is less toxic.

Procedure (according to Tanamati 2005)

Add to 1 g of sample 18 ml of hexane/propan-2-ol (3:2; HIP). Homogenize the mixture for 30 sec and filter the suspension by using a sintered glass Büchner funnel (with a filter paper of 14 μm) fitted with a ball joint for use under pressure. Wash three times the homogenizer, funnel and residue, with 2-ml portions of HIP; re-suspend the residue each time and allow soaking for 2 min just before applying air pressure. Remove the extract by mixing the pooled filtrates for 1 min with 12 ml of aqueous sodium sulphate solution (prepared from 1 g of anhydrous salt and 15 ml of water). The two layers formed in the process represent a volume of 18 ml. After processing the sample as described above, the lipid is isolated in the upper phase (hexane-rich layer).

The Smedes Method (BDS)

Smedes (1999) proposed a modified Bligh and Dyer method, using methanol instead of propan2-ol and chloroform instead of cyclohexane. This modification can be explained by the lower density of cyclohexane, which subsequently separates on top of the extraction mixture.

Procedure (according to Tanamati 2005)

Weigh a sample containing <1 g of lipid and <5 g of water to a 100-ml glass jar. Add 16 ml of propan-2-ol and 20 ml cyclohexane and mix for 2 min using an UltraTurrax mixer. Taking into account the amount of water in the original sample, more water is included to obtain 22 g of water and mixed for 1 min. Separate the phases by centrifugation (450 × g), and transfer the organic phase to an evaporation flask using a glass pipette. In a second extraction process, add 20 ml of cyclohexane and 2.6 ml of propan-2-ol and mix for 1 min in an UltraTurrax mixer. Combine the organic phase with the first extract and evaporate the solvent on a water bath at 85 °C. Transfer quantitatively the residue to a wide-mouthed weighing flask, with one glass petri dish using a few millilitres of the cyclohexane/propan-2-ol mixture or diethyl ether. Allow the solvent to evaporate to dryness in a water bath (ca. 50 °C), and place the resulting material in an oven at 103 °C for 1 h. The lipid content is determined by difference.

The Jensen Method

Jensen et al. (1972) proposed a modified Folch method, using only non-halogenated solvents. After that, Jensen et al. (2003) improved the original Jensen method by investigating other non-chlorinated solvent mixtures that may result in quantitative lipid recoveries when applied to organisms with very low lipid content.

Procedure (according to Jensen et al. 2003)

Weigh 10 g of sample, transfer to an upper funnel (Fig. 6.1) and homogenize in an Ultra Turrax for 1min with solvent mixture containing 25 ml of acetone and 10 ml of normal hexane. Transfer to a lower funnel (which contains 50 ml of 0.1 M phosphoric acid in an aqueous 0.9% sodium chloride solution) through a filter by exerting gentle pressure from compressed nitrogen. Homogenize the sample remaining in the top funnel for 1 min with 25 ml of a mixture with normal hexane/diethyl ether 9:1. Transfer to the lower funnel as described above. Finally, add 25 ml of a mixture of normal hexane/diethyl ether 9:1, shake or stir the sample with a glass rod, and transfer also to the lower funnel. During homogenization, small particles of tissue may be collected in the liquid present between the filter plate and the tap. To prevent this material from passing into the lower funnel, a small portion of the

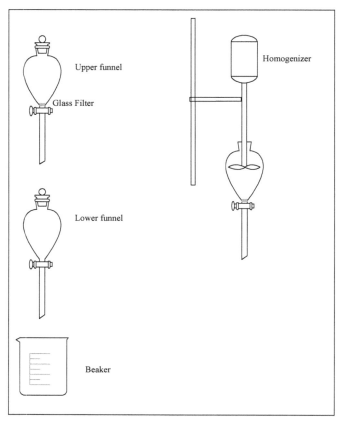

Figure 6.1 Glass apparatus used in the Jensen method (adapted from: Jensen et al. 2003).

extract should be first tapped off into a test tube without nitrogen pressure until a clear phase appears and this can be passed back to the top funnel. To avoid formation of an emulsion, the lower funnel cannot be shaked but just sealed and turned upside down 30 times. After phase separation, transfer the lower aqueous phase to a 100 ml beaker. To avoid water in the organic extract, it is necessary to rotate the lower funnel and transfer any additional water into the beaker. Decant the organic phase into a pre-weighed beaker. Finally, return the aqueous phase to the lower funnel and re-extract with 15 ml of hexane: diethyl ether (9:1 v/v) as above. Evaporate the combined organic phases in the beaker in a cupboard overnight.

Schmid-Bondzynski-Ratzlaff Method

The Schmidt-Bondzynski-Ratzlaff method calls for acid digestion before liquid-liquid extraction of the sample. This method can be applied if the quantification of total lipids defined as the sum of the free and bound lipids, both amphipathic and nonpolar, is the final goal.

This method is commonly applied for determining the lipid content of dairy products; however it can also be applied for meat products.

Procedure (according to Tanamati 2005)

Weigh approximately 3–5 g of homogenized sample on aluminium foil and transfer to an extraction tube. Add 10 ml of hydrochloric acid (8 M) and place the tube in a boiling water bath for 1 h. After cooling the sample to approximately 30°C, add 10 ml of 95% (v/v) ethanol and mix the sample. Add diethyl ether (25 ml), mix, and add another 25 ml of petroleum ether and mix again. Allow the tube to stand overnight to effect phase separation. Siphon off the ether phase into a flat-bottomed flask, and re-extract the sample with 30 ml of a diethyl ether/petroleum ether (50:50, vol/vol) mixture. After phase separation, siphon the organic phase into the same flask as noted previously. Repeat the extraction a third time, and collect the organic phases in the same flask. Evaporate the solvent in the flask, and place the flask in a drying oven for 2 h at 102–105°C. Finally, weigh the fat in the flask.

6.2.3 Solid-Liquid Extractions

The Soxhlet Method

Lipid contents may also be estimated using Soxhlet method which is the officially recommended method (AOAC 2005). Despite being the recommended method, some drawbacks can be pointed out: long time required for extraction and the large volume of solvent used, which is not only expensive, but also can be harmful to the health and to the environment; incomplete extraction of amphipathic lipids into nonpolar solvents for lipid classification, which may not allow for an accurate determination.

Procedure (according to AOAC 2005)

Add 1 g of homogenised sample and place 2 g of sodium sulphate in a Soxhlet extractor. Let 30 ml of n-hexane pass through, in hot extraction mode at 140 °C for 4 h. Remove the extraction container and then evaporate the solvent in an oven at 104 °C, until constant weight.

6.2.4 Lipid Extraction with Nonorganic Solvents

Supercritical Fluid Extraction (SFE)

This technique has been used for the extraction of lipids from various matrices, including dehydrated foods, meats (Berg et al. 2002), oilseeds and fried foods and is becoming a standard method for the preparation and analysis of lipid-containing sample matrices. Recently SFE incorporating supercritical carbon dioxide (SC-CO_2) as the extracting agent has been approved as the AOCS official method for measuring oil content.

Besides the widespread use, results can be compromised if an improper sample matrix preparation prior to extraction is applied, or if an inadequate extract collection method is chosen. Furthermore, the characteristics of the sample matrix should be considered. Lipid content and moisture content of the samples greatly affect the extraction efficiency of lipids. Polar lipids such as phospholipids have sparing solubility in SC-CO_2. In this case, the addition of an entrainer or modifier such as ethanol should be applied in order to improve the extraction (King 2002). A high moisture level in the sample matrix can also decrease the contact between sample and SC-CO_2 as well as the diffusion of lipids out of the sample. Thus, lyophilisation could be necessary for wet samples, such as meat and fish (Shahidi and Wanasundra 2008).

Procedure (according to Berg et al. 1997 and Zhou et al. 2010)

Weigh 5 g (e.g., meat) to 30 g (e.g., scallop) of homogenized sample and mix for 1 min with 1 g hydromatrix using a glass pestle and mortar. Transfer the mixture to an extraction vessel. Prepare the extraction equipment. Use carbon dioxide as the cryogenic gas, required for cooling different zones in the SFE apparatus and as the extracting medium, in most cases together with 95% ethanol as modifier (0–10%). Regulate the flow rate of the carbon dioxide. Set the pressure and temperature at optimal conditions. The parameters of the supercritical CO_2 extraction should be chosen according to previous studies (e.g., Ma et al. 2006; Sahena et al. 2010). After finishing the extraction, collect the lipid, weigh, and store at 20 °C until further treatment. If needed, evaporate the solvent in the vials with compressed air and dry the fat at room temperature overnight in a fume hood. Weighing of the vials before extraction and after solvent evaporation gives the total fat content of the sample.

Microwave-Assisted Extraction (MAE)

Microwave-assisted extraction (MAE) was developed by Batista et al. (2001) to extract lipids from fish for the determination of the fatty acid composition. The equipment consists of an open-vessel extraction apparatus similar to the system of Soxhlet. Microwave energy is used to heat solvents in contact with solid samples and to partition compounds of interest from the sample into the solvent. The method is suitable for the extraction of wet materials.

Procedure (according to Ramalhosa et al. 2012)

Weigh 1 g of homogenized sample and 2 g of sodium sulphate for a quartz extraction vessel of a Microwave Accelerated Reaction System, MARS-X, 1500W (CEM, Mathews, NC, USA), configured with a 14 position carousel (Santorius BasicPlus, P 211D, Germany). Monitor during the operation both the temperature and pressure in a single vessel. Use a magnetic stirrer in each extraction vessel and a sensor to register the solvent leaks in the interior of the microwave oven. After adding to each sample 30 ml of petroleum ether:acetone (2:1, v/v), close the vessels.

The operational parameters should be the following:
- magnetron power 100%;
- time to reach settings 10 min;
- extraction duration 20 min;
- medium speed stirring;
- extraction temperature 90 °C and maximum vessel pressure cut off 1.38×10^6 Pa.

After cooling, filter the extract through Whatman GF/C filter paper, fill with sodium sulphate and collect into an evaporator tube. Evaporate the extract in a rotary evaporator at 45 °C. Dry the residue in an oven at 104 °C for 1 h.

Note: The optimum extraction conditions were based on Ramalhosa et al. 2012 concerning fat composition of the most consumed fish species in Portugal.

6.3 Analysis of Lipid Extracts from Fish and Meat Samples

Lipids in fish and meat samples may be characterized by assessing the amount of fatty acids, mono-, di-, and triacylglycerols, phospholipids, sterols, lipid-soluble pigments and vitamins. Another approach for categorizing lipid fractions regards their nutritive value, which comprises the measurement of not only total fat, but also requires quantification of saturated fat, monounsaturated fat, polyunsaturated fatty acids and *trans* isomer fatty acids.

Classical analytical procedures for lipid analysis are essentially chemical in nature but increasingly they have been replaced by instrumental techniques, based essentially on physical properties. Instrumental methods, such as chromatography and spectroscopy, are generally faster, more accurate, less laborious and require fewer material although they are more expensive. Analysis of lipid extracts from fish and meat samples uses a wide diversity of methods. This sub-chapter highlights only some of the most representative, mainly regarding triacylglycerol and fatty acid composition analysis.

6.3.1 Classical Analytical Procedures

A number of chemical methods have been established to provide information about the type of lipids present in edible fats and oils. Most of these techniques only give information about the average properties of the lipid components present, e.g., the average molecular weight, degree of unsaturation or amount of acids present. Nevertheless, they are easy to perform and do not require expensive equipment, and therefore they are widely used.

Degree of Unsaturation

The iodine value (IV) provides a measure of the average degree of unsaturation of a lipid and is expressed as the number of grams of iodine absorbed per 100 grams of sample. The higher the amount of unsaturation, the more iodine is absorbed and the higher the iodine value (Knothe 2002). Typical iodine values of meat and fish samples are shown in Table 6.2.

Table 6.2 Typical iodine values of meat fat (Meat Research Corporation 1998) and fish oils (from Endo et al. 2005).

Samples	Iodine value
Meat Fat	
Beef tallow	42–48
Mutton tallow	32–44
Pork lard	50–65
Poultry fat	50–80
Fish oils	
Salmon	165.8
Sardine	156.2
Tuna	162.0
Cod	142.0
Squid	168.0

The higher IV attained for fish samples indicate that they are more susceptible to oxidation than meat due to the higher degree of unsaturation.

Determination of IV gives a reasonable estimate of lipid unsaturation unless the double bonds are not conjugated with each other or with carbonyl oxygen. Moreover, the assay should be carried out in the absence of light and with an excess of halogen reagent (Allen 1955).

Although many different iodine value procedures have been developed, the most commonly used is the Wijs method. The standard Wijs method is a titration assay based on the reaction between a known excess of iodine monochlorine and the double bonds in the unsaturated lipids. The remaining iodine monochlorine is measured by adding excess potassium iodide to the solution to liberate iodine and then titrating with a sodium thiosulfate solution in the presence of starch to assess the concentration of iodine released. The lipid to be analysed is weighed and dissolved in a suitable solvent. The employed solvent in the standard method is carbon tetrachloride but the recognition of its potential carcinogenic risk lead to its replacement by a mixture of cyclohexane and glacial acetic acid (Pocklington 1990).

The iodine value can also be estimated from a fatty acid profile analysis assessed by gas chromatography (Ham et al. 1998). Recently, near-infrared spectroscopy showed to be an effective technique for determining the IV value of fish oils (Endo et al. 2005).

170 Methods in Food Analysis

Protocol 1: Wijs method (according to Wrolstad et al. 2005 and to AOCS official method Cd 1–25, 1993)

Wijs solution

Dissolve 16.2 g iodine monochloride in a 1 L. volumetric flask with glacial acetic acid. Store the Wijs solution in an amber bottle with paraffin until it is used. Wijs solution is sensitive to temperature, moisture and light.

Potassium iodide solution

Dissolve 15 g of potassium iodide in 100 ml distilled water in a volumetric flask. Transfer to an amber glass bottle and store in a cool, dark place. Potassium iodide is light sensitive and should be discarded if it turns brown.

Sodium thiosulfate standard solution, 0.1 N

First, dry potassium dichromate for 2 h in an oven at 100 °C. Cool down in a desiccator and accurately weigh between 0.2 and 0.23 g of potassium dichromate. Simultaneously, weigh 24.82 g of sodium thiosulfate pentahydrate and transfer to a 1 L. volumetric flask, making up the volume with distilled water. Dissolve potassium dichromate in 80 ml of distilled water. Then add 2 g of potassium iodide and 8 ml of concentrated hydrochloric acid. Mix properly and titrate with the sodium thiosulfate solution until the brown colour just changes to yellowish-green. Add 1 ml of soluble starch solution and titrate until blue colour turns light green. Record the amount of sodium thiosulfate consumed in the titration assay. Estimate the normality of the sodium thiosulfate solution according to the following equation:

$$Normality\ of\ sodium\ thiosulfate\ = \frac{g\ K_2Cr_2O_7 \times 1000}{mL\ Na_2S_2O_3 \times 49.032} \qquad \text{eq. (6.1)}$$

Soluble starch solution, 0.5%

Put about 150 ml of distilled water in a 500 ml beaker. Heat the water, stirring regularly, until it gets very warm to the touch. Slowly mix in 1 g of soluble starch. Take the solution to a gentle boil until the solution becomes clear. Cool down and transfer to a 200 ml volumetric flask. Make up the volume. Transfer to a storage bottle. Store at 4 °C and check periodically for signs of contamination.

Procedure

If the sample to be analysed is a fat, it must be melted at a temperature that does not exceed 10 °C above the melting point. Then, filter the sample through a Whatman n°1 or n°4 filter. If there is water in the sample, remove it by adding anhydrous sodium sulphate to the oil and afterwards filter the sample. Cool to 68 ± 2 °C in a desiccator. At the end, the sample should be mixed to become properly homogenized. Weigh the required amount of sample to an accuracy of 0.001 g (see Table 6.3). Place the sample into a 500 ml Erlenmeyer flask with 24/40 ground-glass joint. Add 20 ml of cyclohexane washing down any adhering sample from the neck or sides of the flask. Swirl the flask to ensure that the sample is completely dissolved. Dispense 25 ml of Wijs solution into the flask, swirl to mix, then stopper immediately and place in a cool (25 °C to 30 °C), dark place for 1 or 2 hours. Simultaneously, prepare a blank using all the reagents except the lipid sample. Remove the flask from storage and promptly dispense 20 ml of potassium iodide solution. Swirl the flask and stop the reaction by adding 100 ml of distilled water. Titrate, with vigorous shaking/stirring, with 0.1 N sodium thiosulfate using a 50 ml burette. The blank flask should be titrated first followed by the sample flask. When the solution turns to pale yellow stop the titration. Then, add 1 to 2 ml soluble starch indicator and proceed with the titration until the blue/brown colour disappears. Record the volume of sodium thiosulfate that was used.

Calculation

$$\text{iodine value} = \frac{(B-S) \times N \times 126.9}{W} \times 100 \qquad \text{eq. (6.2)}$$

where:
iodine value is expressed as g iodine absorbed per 100 g of sample
B is the volume of titrant (ml) for blank
S is the volume of titrant (ml) for sample
N is the normality of sodium thiosulfate solution (mol/1000 ml)
126.9 is the molar weight of iodine (g/mol)
W is the sample mass (g)

Table 6.3 Sample weights to be used to determine the iodine value.

Expected iodine value	Sample weight (g)
5	3.000
5–20	1.000
21–50	0.400
51–100	0.200
101–150	0.130
151–200	0.100

Free Fatty Acid Content

Free fatty acids (FFA) are end products of the hydrolysis reaction of ester bonds in lipids catalysed by lipases (Murty et al. 2002) or by the presence of water. Partial hydrolysis of triacylglycerol will yield mono- and diacylglycerols and free fatty acids. When hydrolysis is carried out to completion the triacylglycerol will hydrolyse to yield glycerol and free fatty acids.

The accumulation of FFA during the processing or storage of foodstuffs may compromise the quality and the shelf-life of the final product. Their presence mainly causes textural alterations due to their association with proteins. FFA content has been used to establish the grade of deterioration in meat (Kanatt et al. 1998; Chukwu and Imodiboh 2009) and fish samples (Roldán et al. 1985; de Koning and Mol 1991).

Several methods have been implemented to evaluate the FFA content. The standard procedure is based on a titration assay of a liquid fat sample, to which neutralized 95% ethanol and phenolphthalein indicator are added, with a sodium hydroxide aqueous solution.

Protocol 2: Titrimetric method for determination of FFA content—standard procedure (according to Wrolstad et al. 2005 and to AOCS Official method Ca5a-40, 1993)

Neutralized 95% ethanol

Titrate 95% ethanol with a few drops of standardized alkali in the presence of phenolphthalein indicator. Proceed until a faint pink colour persists for 15 to 30 sec.

Phenolphthalein solution

Weigh 1 g of phenolphthalein and transfer to a 100 ml volumetric flask containing 50 ml of 95% ethanol. Stop and shake vigorously for a few minutes. Add 20 ml ethanol and shake until phenolphthalein is in solution. Bring the volume to 100 ml. Store in a glass bottle.

Procedure

Ensure that the sample is well mixed and entirely liquid before weighing. Select from Table 6.4 the sample size, amount of alcohol and normality of sodium hydroxide to be used based on the predicted FFA level. Weigh the sample (with an accuracy of 0.001 g) into a 250 ml Erlenmeyer flask. Add the proper volume of neutralized 95% alcohol preheated to 60 or 65 °C. After this, add 1 ml of phenolphthalein solution as the titration indicator. Swirl to ensure that the sample is well homogenized. Titrate immediately with sodium hydroxide solution until a pale pink colour is observed. Record the amount of sodium hydroxide solution used.

Table 6.4. Sample weights and alcohol volumes for determination of FFA content in fats and oils.

Expected FFA (%)	Sample weight (g)	Amount of alcohol (ml)	Normality of NaOH
0.0–0.2	56.4 ± 0.20	50	0.10
0.2–1.0	28.2 ± 0.20	50	0.10
1.0–30.0	7.05 ± 0.05	75	0.25
30.0–50.0	7.05 ± 0.05	100	0.25 or 1.0
50.0–100.0	3.525 ± 0.001	100	1.0

Calculation:

$$\% \, FFA \, (as \, oleic \, acid) = \frac{V \times N \times 282}{W} \times 100 \qquad \text{eq. (6.3)}$$

where:
% FFA is the percentage of free fatty acid (g/100 g) expressed as oleic acid;
V is the volume of sodium hydroxide titrant (ml);
N is the normality of sodium hydroxide titrant (mol/1000 ml);
282 is the molecular weight of oleic acid (g/mol);
W is the sample mass (g).

Small adjustments have been proposed to this method according to the expected quantities of FFA and to the nature and amount of sample available. As an example, see the references (de Koning and Mol 1991; de Koning 1999; Bernárdez et al. 2005) for the analysis of fish samples and (Kanatt et al. 1998; Chukwu and Imodiboh 2009) meat samples.

Ke and Woyewoda (1978) proposed a titrimetric method to evaluate the FFA content in various lipids, mostly in fish products, even when the samples are highly coloured or rancid.

Protocol 3: Modified titrimetric method for the determination of FFA content in tissues (according to Ke and Woyewoda 1978).

Procedure

Accurately weigh 1 g of lipid sample into a 250 ml Erlenmeyer flask. Dissolve in 50 ml of chloroform. Then add a few drops of *m*-cresol indicator. Titrate the solution with aqueous 0.05 M hydroxide solution until the purple endpoint is reached. Simultaneously, prepare a blank titration, i.e., without the lipid sample.

Colorimetric methods have also been used to determine the FFA content. One of the most common procedures for the analysis of FFA in fish samples is the Lowry and Tinsley method (Lowry and Tinsley 1976). This method includes oil dissolution in benzene then allowing the FFA to react with a cupric acetate solution. The organic solvent turns to a blue colour due to the FFA–cupric ion complex, which has a maximum absorbance between 640 and 690 nm. This rapid and accurate method has the limitation of using benzene that is a carcinogenic solvent. In order to develop a safer and economical method, Bernárdez et al. (2005) implemented some modifications to the Lowry and Tinsley method which are stated in protocol 4.

Fourier Transform IR (FTIR) spectroscopy can also be used to evaluate the FFA content since there is a band attributed to the carboxyl group (COOH) in the central region of the mid-IR spectrum (Sherazi et al. 2007).

Protocol 4: Lowry and Tinsley Method (1976)

Cupric acetate-pyridine reagent 5% (w/v)

Weigh 5 g of cupric acetate and transfer to a 100 ml volumetric flask, making up the volume with distilled water. Adjust the pH to 6.0 – 6.2 using pyridine.

Procedure

Remove any solvent present in the sample using a nitrogen stream. Place a sample portion containing 2.0–14.0 µmol of FFA in a screw cap culture tube. Accurately add 5.0 ml of benzene and swirl to dissolve the sample. Slight warming may be necessary to effect dissolution. Add 1.0 ml of cupric acetate-pyridine reagent and shake the biphasic system for 2 min. Centrifuge at 1470 g for 5 min. Read the upper layer in the spectrophotometer at 715 nm. The FFA concentration in the sample may be calculated as micromolar oleic acid based on a standard curve.

Modification to protocol 4 (according to Bernárdez et al. 2005)

Instead of 5 ml of benzene add 3 ml of cyclohexane. Let the contact time between the two phases be only 30 s. Read the upper layer at 710 nm.

Saponification Value

The saponifiable fraction includes the derivatives of fatty acids, mainly triacylglycerol (as well as mono- and diacylglycerols), phospholipids, glycolipids, waxes and sterol esters.

Saponification is the process of breaking down or degrading a neutral fat into glycerol and fatty acids by treatment with alkali.

The saponification number is a measure of the average molecular weight of the triacylglycerol in a sample. It is defined as the milligrams of potassium hydroxide required to saponify 1 g of fat, that is, to neutralize the free fatty acids (Pomeranz and Meloan 1994). The smaller the saponification value, the larger the average molecular weight of the triacylglycerols present, i.e., saponification value is inversely proportional to the mean molecular weight of fatty acids (or chain length).

The standard method involves the use of excess alcoholic potassium hydroxide, which catalyses the saponification/release of the free fatty acids from the glycerol backbone.

The alkali required for saponification is determined by titration of the excess potassium hydroxide with standard hydrochloric acid using phenolphthalein as the indicator. The amount and normality of hydrochloric acid used for neutralization can then be used to calculate the saponification value.

Protocol 5: Determination of saponification value—standard procedure (according to the AOAC official method 920.160, 2000)

Alcoholic potassium hydroxide solution

Reflux 1.2 l alcohol for 30 min with 10 g potassium hydroxide and 6 g granulated aluminium. Distil and collect 1 l alcohol after discarding the first 50 ml. Dissolve 40 g potassium hydroxide in this alcohol, keeping temperature below 15 °C. Allow to stand overnight. Decant the clear liquid and keep it in a tightly closed bottle.

Procedure

Melt the sample and pass it through a filter paper. Ensure that the sample is properly dried. Homogenize the sample and weigh about 1.5 to 2.0 g into a 250 ml Erlenmeyer flask. Pipette 25 ml of alcoholic potassium hydroxide solution into the flask. Conduct a blank determination along with the sample. Attach a suitable condenser to the sample flask and heat it in a water bath. Adopt the same procedure for the blank. Allow the sample to gently reflux until it is completely saponified (usually it takes 1 hour). The completeness of saponification is indicated by a clear, homogeneous sample. Once saponification is complete, remove from heat and let the sample cool while still attached to the condenser. Once the sample has cooled, disconnect it from the condenser and add 1 ml phenolphthalein solution. Titrate with the standardized hydrochloric acid 0.5 N until a pink colour disappears, indicating the end point. Record the amount of hydrochloric acid used.

Calculation:

$$saponification\ value = \frac{(B-S) \times N \times 56.1}{W} \qquad \text{eq. (6.4)}$$

where:
saponification value is expressed as mg potassium hydroxide per g of sample
B is the volume of standard hydrochloric acid (ml) required for the blank
S is the volume of standard hydrochloric acid (ml) required for the sample
N is the normality of the standard hydrochloric acid (mmol/ml)
56.1 is the molar weight of potassium hydroxide (mg/mmol)
W is the sample mass (g)

Measurement of Oxidative Deterioration

As previously mentioned, lipids are prone to oxidation when exposed to light, heat, enzymes and metals, leading to nutritional losses and the formation of off-flavours (Ladikos and Lougovois 1990; Min and Ahn 2005; Barriuso et al. 2013). Several analytical methods are available for the measurement of lipid oxidation in foods (Shahidi and Zhong 2005; Antolovich et al. 2002 ; Barriuso et al. 2013). Changes in chemical, physical, or organoleptic properties of fats and oils may be monitored to assess the extent of lipid oxidation. However, there is no uniform and standard procedure for detecting all oxidative changes in all food systems. The available methods to monitor lipid oxidation in foods and biological systems may be divided into two groups. The first group measures primary oxidative changes and the second determines secondary changes that occur in each system.

Measurement of Primary Products of Oxidation

As previously referred, hydroperoxides are primary oxidation products that are continuously formed during lipid oxidation and easily decomposed into several non-volatile and volatile secondary products. At the initial stage of oxidation, the formation rate of hydroperoxides prevails over their decomposition rate but this tendency is reversed at later stages (Dobarganes and Velasco 2002).

The Peroxide value (PV) represents the total hydroperoxide content and is commonly expressed as the milliequivalents (mEq) of peroxides per kilogram of a sample or as millimolar of peroxide per kilogram of lipid. Several analytical procedures are available for the determination of PV in fish and meat samples, namely the iodometric titration, the ferric ion complex measurement spectrophotometry and the infrared spectroscopy (Shahidi and Zhong 2005; Barriuso et al. 2013).

The iodometric titration method is based on the oxidation of the iodine ion by hydroperoxides followed by the measurement of the liberated iodine by titration with a standard sodium thiosulfate solution using starch as an endpoint indicator (Barriuso et al. 2013).

Although iodometric titration is the most widely used method for measurement of PV, it is time consuming, laborious and requires a large amount of sample. Furthermore, oxygen in the air, light and absorption of iodine by the unsaturated fatty acids in the sample may interfere with the results (Sun et al. 2011). Other drawbacks are the lack of sensitivity and the difficulties in determining the titration end point (Shahidi and Zhong 2005).

Protocol 6: Iodometric Titration Method (according to Wrolstad et al. 2005 and to the AOCS Official method Cd 8–53, 1998)

Procedure

Start by removing air from the solutions. Select from Table 6.5 the sample size according to the expected PV. Weight the sample with an accuracy of ± 0.05 g into a 250 ml glass stoppered Erlenmeyer flask. Add 30 ml of 3:2 acetic acid/chloroform solution and swirl to achieve complete sample dissolution. Then add 0.5 ml of saturated potassium iodide solution to the flask. Allow the solution to stand (for 1 min, precisely) with occasional stirring. After add 30 ml of distilled water. Fill a 10 or 25 ml graduated burette with 0.1 N standardized sodium thiosulfate solution and add it gradually to the flask to titrate, with vigorous shaking/stirring, until the yellow color has almost disappeared. Add 0.5 ml of 1% (w/v) starch indicator solution. Proceed with the titration assay, adding sodium thiosulfate dropwise, until the violet color disappears. Register the amount of titrant used. Simultaneously, prepare a blank using all of the reagents except the lipid sample.

Table 6.5 Sample size for iodometric PV determination.

Expected PV (meq active oxygen/kg sample)	Weight of sample (g)
0–2	5
2–10	2
10–25	1
25–50	0.5
50–100	0.3

Calculation:

$$\text{peroxide value} = \frac{(S-B) \times N}{W} \times 1000 \qquad \text{eq. (6.5)}$$

where:
peroxide value is expressed as mEq peroxide per kg of sample
S is the volume of titrant (ml) for sample
B is the volume of titrant (ml) for blank

N is the normality of sodium thiosulfate solution (mEq/ml)
W is the sample mass (g)

Several colorimetric methods for determination of PV have been reported (Shantha and Decker 1994; Shahidi and Zhong 2005; Barriuso et al. 2013). They are based on the ability of hydroperoxides to oxidize ferrous ions to ferric ion which are complexed by either thiocyanate or xylenol orange producing chromophores that can be measured spectrophotometrically. Ferric thiocyanate is a red complex with an absorption maximum of 500 nm. The ferrous oxidation of xylenol orange (FOX) procedure uses dye xylenol orange to form blue purple complex with a maximum absorption at 550–600 nm. Both methods present several advantages in comparison to the iodometric method (Table 6.6).

The PV can also be assessed by a rapid method based on flow analysis with Fourier transform IR spectroscopy (Ruíz et al. 2001).

Protocol 7: The ferrous oxidation of xylenol orange (FOX) method (according to Wrolstad et al. 2005)

Procedure

Accurately weigh 0.01 to 0.30 g of oil or lipid extract sample into a 16 x125-mm borosilicate glass tube. Pipette 9.9 ml of 7:3 chloroform/methanol solution into the same tube. Vortex sample for 2 to 4 sec. Add 50 µl of 10 mM xylenol orange solution to the sample, vortex for 2 to 4 sec, and then add 50 µl iron(II) chloride solution and vortex again. Allow solution to stand precisely 5 min at room temperature. After this, determine its absorbance at 560 nm. Set the spectrophotometer zero with chloroform/methanol solution. Construct a standard curve by adding to a series of 16 x 125-mm borosilicate glass varying aliquots of an iron (III)-chloride standard solution (10 µl/ml), 50 µl of 10 mM xylenol orange solution and complete to a final volume of 10 ml with 7:3 (v/v) chloroform/methanol solution. Suitable amounts of the iron (III) chloride standard solution should range from 0 to 2 ml. The absorbance reading for the test tube with only xylenol orange and the 7:3 chloroform/methanol solution represents the absorbance of the blank. Plot absorbance of the standards *versus* µg Fe^{3+} and fit with a regression line. Only the range between 5 and 20 µg Fe^{3+} should be used.

Calculation:

$$PV = \frac{[(A_S - A_B)] \times m_i}{W \times 55.84 \times 2} \qquad \text{eq. (6.6)}$$

Table 6.6 Summary of methods for analysis of primary oxidation products (adapted from Shahidi and Zhong 2005).

Method	Principle	Advantage	Disadvantage	Studies - Fish Samples	Studies - Meat Samples
Iodometric titration Sensitivity ~ 0.5meq/kg fat	■ Reaction of saturated solution of potassium iodide with hydroperoxides. ■ Titration of the liberated iodine with a standardized solution of sodium thiosulfate using starch as an end point indicator.	■ Standard method	■ Time consuming; ■ Require large amount of sample; ■ Generate a significant amount of waste; ■ Possible absorption of iodine across unsaturated bonds; ■ Possible oxidation of iodine by dissolved oxygen; ■ Difficulties in determining the titration endpoint.	Shantha and Decker 1994 (comparison study) Babalola and Apata 2011 (comparison study)	Shantha and Decker 1994 (comparison study) Babalola and Apata 2011 (comparison study)
Ferric ion complexe-Ferric thiocyanate Sensitivity ~ 0.1meq/kg fat	■ Ferric ion is complexed by thiocyanate leading to the formation of ferric thiocyanate which is a red-violet complex that shows strong absorption at 500–510 nm.	■ Simple ■ Reproducible ■ Sensitive	■ Careful control of the operational conditions is required for accurate measurements.	Shantha and Decker 1994 (comparison study)	Shantha and Decker 1994 (comparison study)
Ferric ion complexe-Ferric xylenol orange Sensitivity ~ 0.5meq/kg fat	■ Ferric ion is complexed by xylenol orange leading to the formation of ferric xylenol orange which is a blue-purple complex with a maximum absorption at 550–600 nm.	■ Rapid ■ Inexpensive ■ Not sensitive to oxygen and light	■ Careful control of the operational conditions is required for accurate measurements.	Eymard and Genot 2003 (modified method) Shantha and Decker, 1994 (comparison study)	Shantha and Decker 1994 (comparison study)

where:

PV is expressed as meq active oxygen/kg sample
A_S is the absorbance of the sample
A_B is the absorbance of the blank
m_i is the inverse of the slope (obtained from calibration)
W is the weight of the sample (g)
55.84 is the atomic weigh of iron

Measurement of Secondary Products of Oxidation

Secondary oxidation products are a suitable index of lipid oxidation due to the fact that they are odor-active and stable compounds, in comparison to primary products (hydroperoxides), which are colourless, flavourless and usually labile compounds (Shahidi and Wanasundara 2008).

Secondary oxidation compounds comprise aldehydes, ketones, hydrocarbons, volatile organic acids, alcohols and epoxy compounds, among others.

One of the most extensively employed methods to detect oxidative deterioration in foodstuffs is the thiobarbituric acid (TBA) test. This procedure is based on the formation of malondialdehyde (MDA), during the autoxidation of polyunsaturated fatty acids, followed by reaction with TBA to form a pink complex that is measured spectrophotometrically at 530–535 nm (Shahidi and Zhong 2005; Barriuso et al. 2013). Reaction kinetics is influenced by the concentration of TBA solution, temperature and the pH (Fernandez et al. 1997). The extent of oxidation is stated as the TBA value and is reported as miligrams of MDA equivalents per kilogram of sample or as micromoles of MDA equivalents per gram of sample.

Nevertheless, TBA is not selective to MDA and can also react with many other compounds such as aldehydes, carbohydrates, amino acids and nucleic acids leading to overestimation and variability in the results attained by the TBA method (Salih et al. 1987). For this reason, this method is also known as thiobarbituric acid reactive substances (TBARS) test.

Several procedures have been described to perform TBA test in fish and meat products namely by (i) direct heating of the sample with TBA, followed by separation of the pink complex produced by centrifugation, (ii) distillation of the sample, followed by the reaction of the distillate with the TBA, (iii) extraction of MDA using aqueous trichloroacetic or perchloric acid and subsequent reaction with TBA and (iv) extraction of the lipid fraction of the sample with organic solvents and reaction of the extract with the TBA (Shahidi and Zhong 2005; Barriuso et al. 2013).

The acid extraction method is the most commonly used method for determining TBA value in fish and meat samples together with the distillation method.

Protocol 8: Thiobarbituric Acid Test (TBA): Acid Extraction Method (according to Papastergiadis et al. 2012)

Procedure

Weigh approximately 7 g of sample in a 50 ml falcon tube. Add 15 ml of 7.5% (w/v) of trichloroacetic acid with 0.1% (w/v) of ethylenediaminetetraacetic acid (EDTA) and 0.1% (w/v) of propyl gallate. Homogenize the mixture properly (for instance, with an Ultraturax, for 1 min at 18 000 rpm). Adjust the volume to 30 ml with addition of trichloroacetic acid. Filter the homogenate through 150 mm filter paper. Transfer 2.5 ml of extract and mix with 2.5 ml of thiobarbituric acid reagent (46 mM in 99% glacial acetic acid) in a test tube. Heat in a boiling water bath for 35 min. Chill the reaction mixture. Measure the absorbance at 532 nm. Prepare standard solutions of malondialdehyde in 7.5% thiobarbituric acid reagent from 1,1,3,3-tetraethoxypropane. The calibration curve should range from 0.6 to 10 µM.

Another common procedure to assess secondary oxidation products is the *p*-anisidine value (*p*-AnV) method which measures the content of aldehydes (mainly 2-alkenals and 2, 4-alkadienals) generated during the decomposition of hydroperoxides. It is based on the reaction, under acidic conditions, between *p*-methoxyaniline (anisidine) and the aldehydic compounds leading to the formation of a yellowish product that absorb at 350 nm (Shahidi and Zhong 2005).

Protocol 9: p-Anisidine Value (p-AnV) (According to AOCS Official method Cd 18–90, 2003)

Procedure

Weigh accurately 0.5 g to 4 g of oil into a 25 ml volumetric flask. Dissolve and make up to volume with isooctane. Measure the absorbance of the solution at 350 nm in a 1 cm glass cell against pure isooctane. Transfer 5 ml of the fat solution into a test tube A and 5 ml isooctane into a test tube B. Add 1ml of 0.25 g/100 ml *p*-anisidine solution in glacial acetic acid to both test tubes. Stopper the tubes, shake vigorously and leave in a dark place for 10 min, precisely. Measure the absorbance of the content of tube A against tube B at 350 nm in a 1 cm glass cell.

Calculation:

$$pAnV = \frac{[25 \times (1.28 \times A_S - A_B)]}{W} \qquad \text{eq. (6.7)}$$

where:
A_S is the absorbance of the oil solution after reaction with the reagent.
A_B is the absorbance of the initial solution.
W is the mass of the sample (g).

Total oxidation, including primary and secondary oxidation products, may be evaluated by the Totox value which is a combination of PV and p-AnV according to the following equation:

$$Totox\ value = 2PV + pAnV \qquad \text{eq. (6.8)}$$

During lipid oxidation there is a tendency for an increase in PV followed by a reduction as hydroperoxide decompose. In fact, PV and pAnV reflect the oxidation level at early and later stages of oxidation reaction, respectively (Shahidi and Zhong 2005).

Preparation of Fatty Acids Methyl Esters

The most used strategy for fatty acid assay on fish and meat samples comprises the extraction of lipids with proper organic solvents, followed by their transesterification and subsequent chromatographic analysis (Liu 1994).

The nutritional value of a food sample strongly depends on its fatty acid profile. Gas chromatography (GC) has been selected as a powerful instrumental technique to determine the fatty acid profile of a lipid sample. A derivatization procedure should be carried out prior to GC analysis in order to increase the volatility of lipid components, thus providing a better resolution of chromatographic peaks (by avoiding peak tailing, peak asymmetry and peak shouldering) as well as a reduction on the time to carry out the analysis (Liu 1994; Brondz 2002).

Although several derivatization assays have been reported in the literature, most of them involve the conversion of fatty acids into their corresponding esters, usually methyl esters.

The conversion of fatty acids, in a complex lipid, into fatty acid methyl esters (FAME) may be attained mainly by two mechanisms: methylation following hydrolysis of the fatty acids or direct transmethylation. The first mechanism comprises saponification (alkaline hydrolysis, see Protocol 5) followed by acid-catalyzed transmethylation/methylation or alkali-catalyzed transmethylation. Direct transmethylation is usually a one-step reaction involving alkaline or acidic catalysts (Liu 1994).

Acid-catalyzed Transmethylation/Methylation

Free fatty acids (FFA) are methylated and O-acyl lipids transmethylated by heating them with a large excess of anhydrous methanol in the presence of an acidic catalyst. Three commonly used acid reagents are hydrogen chloride, sulphuric acid and boron trifluoride.

Anhydrous hydrogen chloride in methanol has been recognized as the mildest reagent. It is simply prepared either by bubbling anhydrous hydrogen chloride gas into methanol or by adding liquid acetyl chloride slowly to methanol. FAME are obtained by heating the reaction mixture in a stoppered tube at 50°C overnight. In order to ensure complete dissolution of non-polar lipid classes, such as cholesterol esters and TAG, an important fraction in meat tissues, a further solvent must be employed. Benzene has often been used for this purpose but, due to its high toxicity, it has been replaced by methylene chloride, toluene and tetrahydrofuran (Raes and De Smet 2009).

A solution of 1%–2% (v/v) concentrated sulfuric acid in methanol is an alternative acid catalyst that has led to similar results as a 5% methanolic hydrogen chloride solution. However, this procedure is not recommended for fish samples, rich in polyunsaturated fatty acids, due to the strong oxidizing nature of sulfuric acid.

Protocol 10: Preparation of fatty acid methyl esters with methanolic sulphuric acid (According to Christie 1989)

Procedure

Weigh up to 50 mg of lipid sample. Dissolve the sample in 1 ml of toluene. Add 2 ml of 1% sulfuric acid in methanol. Heat the reaction mixture in a stoppered tube at 50 °C overnight (16 h) or in reflux for 2 hours. Add 5 ml of water containing sodium chloride (5% (w/v)). Extract the transmethylated fatty acids with hexane 2 x 5 ml, using Pasteur pipettes to separate the layers. Wash the hexane layer with 4 ml of water containing potassium bicarbonate (2% (w/v)). Then, dry it with anhydrous sodium sulfate. Filter the solution. Remove the solvent under reduced pressure in a rotary evaporator or in a stream of nitrogen.

Boron trifluoride in methanol is also used as a transmethylation catalyst and, in particular, as a rapid esterifying reagent for FFA (Morrison and Smith 1964). Nevertheless, several reports stated the appearance of methoxy artifacts probably due to the high concentration usually employed (12% (v/v)) in comparison to other acids (1–5% (v/v)) (Morrison and Smith 1964). Some evidence exists that artifact formation is most likely with old reagents. Also, some reports found losses of fatty acid esters with chain lengths lower than C12. This method is not recommended for lipid samples having conjugated dienoic fatty acids (such as conjugated linoleic acid), namely ruminant meat products.

Protocol 11: Preparation of fatty acid methyl esters with boron trifluoride in methanol after saponification (according to Schlechtriem et al. 2008 and to AOAC Official method 969.33, 1995)

Procedure

Weigh 1 mg of lipid sample. Add 2 ml of methanolic sodium hydroxide solution (0.5 N). Incubate at 100 °C for 10 min. Following saponification, cool down the sample to room temperature. Add 2 ml of boron trifluoride-methanol solution (14% w/w, diluted with water-free methanol to 12.5%). Reheat the sample for 2 min at 100 °C. Cool down the sample once again. Add 1 ml of *n*-heptane. Vortex the sample and reheat to 100 °C for 1 min. Add 1 ml of saturated sodium chloride solution to guarantee the complete phase separation and transfer the heptane phase, containing the methylated fatty acids, to a vial and inject directly into the column of a gas liquid chromatograph.

Base-Catalyzed Transesterification

O-Acyl lipids are ready transesterified in anhydrous methanol in the presence of a basic catalyst. Usually, FFA are not esterified and therefore it is important to ensure anhydrous conditions in the reaction in order to prevent the formation of FFA by hydrolysis.

The widely used reagent has been sodium methoxide (0.5 M) in anhydrous methanol. It is prepared simply by dissolving freshly cleaned sodium in dry methanol. Also, potassium hydroxide in anhydrous methanol, methanolic sodium, and potassium methoxide are often used in concentrations between 0.5 and 2 M. Nevertheless, solvents containing potassium have some drawbacks namely artifact formations, hydrolysis of lipids in presence of trace amounts of water, etc., and therefore are less recommended.

Another base-catalyzed esterification reagent is tetramethylguanidine in methanol (Schuchardt and Lopes 1988). However, this reagent is not appropriate for analysis on meat samples since it is not efficient in the esterification of polar lipids, an important lipid fraction in the meat matrix.

In order to solubilize non-polar lipids, such as cholesterol esters or triacylglycerols, an additional solvent such as toluene or tetrahydrofuran should be employed.

Protocol 12: Preparation of fatty acid methyl esters with sodium methoxide in dry methanol (according to Christie 1989)

Procedure

Weigh up to 50 mg of lipid sample. Dissolve in 1 ml dry toluene in a test tube. Add 2 ml 0.5 M sodium methoxide in anhydrous methanol. Keep the solution at 50°C for 10 min. Add 0.1 ml glacial acetic acid and 5 ml of water.

Reheat the samples for 2 min at 100 °C. Cool down the samples once again. Extract the esters into hexane (2 x 5 ml), using a Pasteur pipette to separate the layers. Dry the hexane layer with anhydrous sodium sulfate and filter. Remove the solvent under reduced pressure on a rotary evaporator.

Several problems may occur during FAME preparation namely (i) incomplete conversion of lipids into FAME, (ii) changes in lipid composition during transesterification, (iii) formation of artefacts, (iv) incomplete extraction of esters into an organic layer and (v) loss of short chain volatiles (Christie 1993; Shantha and Napolitano 1992). According to Li and Watkins (2001), there is no single procedure that works properly in all situations. Therefore, the researcher should know the nature of the sample in order to select the appropriate method.

6.3.2 Instrumental Methods for Lipid Characterization

Thin Layer Chromatography

Thin layer chromatography (TLC) was developed as early as 1938 and has become one of the most popular separation techniques mainly due to its simplicity, reliability and relatively inexpensive equipment used (Shantha and Napolitano 1998).

Several books and reviews report TLC application to lipids for routine separations, identification and quantification (Christie and Han 2010; Fuchs et al. 2011).

The fundamentals of TLC are based on the difference in the affinity of a component toward a stationary and a mobile phase. The most common stationary phases for lipid separations are silica gel, alumina and kieselguhr. Silica gel, undoubtedly the prevalent stationary phase, may be modified by impregnation with other substances in order to enhance the efficiency of separation of certain classes of lipids (Christie and Han 2010; Fuchs et al. 2011).

Quantification has been recognized as the main constraint to the use of TLC. In order to overcome this limitation the TLC/FID Iatroscan system was implemented and has been used routinely for lipid analysis in the last decades (Shantha and Napolitano 1992; Sinanoglou et al. 2013). This technique combines the separation skills of conventional TLC with the quantification power of the flame ionization detector (FID) which is a universal analytical instrument that offers high sensitivity and linearity for carbon-containing organic compounds. TLC/FID has been successfully used for fish (Kaitaranta 1981; Indrasena et al. 2005) and meat analysis (Angelo and James 1993).

Gas Chromatography

Gas chromatography (GC), also named gas-liquid chromatography, is based on the partition of the components of a mixture between an inert gas, as the mobile phase, and a stationary non-volatile liquid phase dispersed on an inert support. The sample is injected into the gas phase where it is volatilised and passed onto the liquid phase, which is held in a column. The components are partitioned between both phases, depending on their relative affinities, and emerge from the end of the column exhibiting peaks of concentration, ideally with a Gaussian distribution. The identification of chromatographic peaks is based on comparison of their retention times, i.e., the elapsed time from injection until the highest-concentration part of the peak has eluted from the column, with those of standards. Commercially available standard mixtures containing accurately known amounts of methyl esters of saturated, monoenoic and polyenoic fatty acids can be purchased and should be analyzed under the same operational conditions as the sample, in order to identify the retention time of the different components of the mixture (Christie 1989; Tranchida et al. 2007).

Gas chromatography is a versatile and accurate technique widely used for the analysis of lipid components. It can be employed namely for evaluating (i) the total fatty acid composition, (ii) the regiodistribution of fatty acids in lipid, (iii) the amount of sterols, (iv) the fat stability and oxidation and (v) the presence of adulterants and antioxidants (Pomeranz and Meloan 1994).

Capillary gas chromatography with flame ionization detector (GC-FID) is the most widely used technique for the analysis of fatty acids, as FAME, in meat and fish samples. Also GC combined with mass spectrometry (MS) is a powerful tool used in identification of compounds.

As previously mentioned, meat and fish samples exhibit distinct fatty acid profiles.

Fish species are generally recognized as excellent dietary sources of HUFA and PUFA, particularly the omega-3 fatty acids EPA and DHA (Abbas et al. 2009; Huynh and Kitts 2009). As an example, Table 6.7 shows the fatty acid profile of three fish species; *Balistes capriscus* (Triggerfish), *Spondyliosoma cantharus* (Black Seabream) and *Labrus bergylta* (Wrasse), abundant in the western coast of Portugal (Simões and Sousa, data not published).

The characteristics of the chromatographic column, namely its length and packing material, influence the resolution of the peaks in the chromatogram. Figure 6.2 is an example of a chromatogram obtained by GC-FID analysis of FAME from a lipid extract of *Balistes capriscus* (Simões and Sousa, data not published).

Table 6.7 Fatty acid profiles of total lipids extracted from *Balistes capriscus*, *Spondyliosoma cantharus* and *Labrus bergylta*. Values are mean value ± SD (% total fatty acids). n.d. stands for "not determined".

Fatty Acid	Balistes capriscus	Spondyliosoma cantharus	Labrus bergylta
C12:0	0.21±0.10	0.06±0.001	0.092±0.05
C14:0	0.27±0.02	1.661±0.05	0.725±0.03
C15:0	n.d.	0.35±0.006	0.290±0.006
C16:0	15.34±0.57	27.80±0.34	22.620±0.06
C16:1 *n*-7	0.86±0.21	0.50±0.003	0.625±0.03
C18:1 *n*-9	26.47±2.40	26.15±0.07	18.245±0.31
C18:2 *n*-6	n.d.	0.54±0.03	0.165±0.007
C18:3 *n*-3	0.60±0.09	n.d.	0.142±0.002
C18:4 *n*-3	n.d.	1.017±0.002	0.165±0.03
C20:1 *n*-9	0.90±0.11	0.77±0.02	0.350±0.01
C20:4 *n*-6	5.67±0.47	2.684±0.006	6.636±0.05
C20:5 *n*-3	5.28±0.36	9.816±0.11	16.131±0.18
C22:5 *n*-3	0.40±0.11	n.d.	0.819±0.004
C22:6 *n*-3	44.30±2.71	28.66±0.44	32.77±0.03

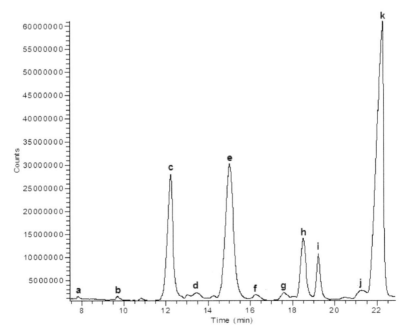

Figure 6.2 Chromatogram of fatty acid methyl esters isolated from *Balistes capriscus*. The peaks stand for: a, C12:0; b, C14:0; c, C16:0; d, C 16:1; e, C18:1; f, C 18:3; g, C 20:1; h, C 20:4; i, C 20:5; j, C 22:5; k, C 22:6.

The operational conditions selected for oven, detector and injector in the chromatograph also affect peak resolution. Table 6.8 summarizes some chromatographic conditions for the analysis of fish samples.

Regarding meat lipids, they are characterized by high contents of SFA and trans-fatty acids (TFA) as well as low levels of PUFA. In particular, the ruminant products are characterized by fatty acids derived from microorganisms present in the rumen (e.g., odd- and branched-chain fatty acids, TFA and CLA), while monogastric products have a much simpler fatty acid profile (mostly linear even fatty acids). Table 6.9 summarizes some chromatographic conditions for the analysis of meat samples.

Headspace solid phase microextraction (HS-SPME) technique coupled with GC has been widely used for measure flavour and off-flavour compounds (e.g., hexanal and heptanal), produced by lipid oxidation in food samples. Several reviews concerning this powerful technique have been published (Kataoka et al. 2000; Wilkes et al. 2000; Pillonel et al. 2002) as well as research work focusing on meat (Estévez et al. 2003; Ganhão et al. 2010) and fish samples (Lee et al. 2003; Miyasaki et al. 2011).

High Performance Liquid Chromatography

High performance liquid chromatography (HPLC) has been recognized as a powerful technique for the analysis of triacylglycerols (TAG), the main constituent of food lipids. Excellent resolution of TAG has been achieved by reversed-phase HPLC (RP-HPLC), which employs a non-polar stationary phase and a polar mobile phase. With relation to the stationary phase, the most commonly used is the octadecylsilyl (C18 or ODS) type, chemically bonded to silica. Acetonitrile is used as the main component of the mobile phase. Other solvents such as acetone, dichloromethane or tetrahydrofuran are added to acetonitrile in proportions that provide high peak resolution. The partition number (PN) influences the elution order of the sample components in the chromatographic column. PN is calculated as $CN - 2DB$, where CN is the carbon number and DB stands for double bond number (Christie 1993; Tranchida et al. 2007).

A recent study reports the composition of TAG for the differentiation of lard and other animal fats (beef, mutton and chicken fats), as well as cod liver oil, using HPLC with refractive index detector (Rohman et al. 2012).

TAG analysis can also be performed by Silver-ion HPLC (Ag^+-HPLC). In this case, separation is based on the DB number, geometrical configuration and position. The separation mechanism comprises the establishment of temporary complexes between TAG DB π-electrons and the silver ions, which act as electron acceptors. Ag^+-HPLC is very efficient in separating lipids differing in the number and positions of DBs (Tranchida et al. 2007).

Table 6.8 Gas chromatographic analysis of fatty acid methyl esters from fish samples.

Analysed Sample	Chromatographic capillary column	Oven Temperature	Detector	Injector	Carrier gas
Twelve Mediterranean Fish Species (Nevigato et al. 2012)	SPB™ PUFA fused silica 30 m x 0.25 mm x 0.20 μm	Initial T=50 °C, 2 min hold, increasing to 210 °C at a rate of 9 °C/min, 40 min hold	FID T= 275 °C	Split mode (1:50) T = 200 °C Injection volume = 2 μl	Helium used in ramp flow mode: 0.8 ml/min for 20 min, then increasing to 1 ml/min at a rate of 0.2 ml/min, 20 min hold, then increasing to 1.2 ml/min at a rate of 0.2 ml/min until the end of the analysis.
Cod liver oil and a marine microalga, Haptophyceae *Pavlova* (Carvalho and Malcata 2005)	Supelcowax-10 60 m x 0.32 mm x 0.25 μm	T=170 °C to 220 °C at a rate of 1 °C/min	FID T= 270 °C	Split mode (for oil samples) Splitless mode (for microalgal samples) T = 250 °C	Helium
Sharpsnout sea bream (Orban et al. 2000)	SPB™ PUFA fused silica 30 m x 0.25 mm x 0.20 μm	T=50°C to 180 °C at a rate of 10 °C/min, 1 min hold, then increasing to 210 °C at a rate of 4 °C/min, hold for 50 min	FID T= 250 °C	Split mode (50:1) T = 200 °C Injection volume = 2 μL	Helium Flow rate = 3.6 ml/min
10 species of marine fish in Malaya (Osman et al. 2001)	SGE BPX-70 60 m x 0.25 mm x0.25 μm	Isothermally at 200 °C	FID T= 260 °C	Split mode (100:1) T = 250 °C	Helium Flow rate = 1.7 ml/min

Table 6.9. Gas chromatographic analysis of fatty acid methyl esters from meat samples.

Analysed Sample	Chromatographic capillary column	Oven Temperature	Detector	Injector	Carrier gas
Pig meat after probiotic administration (Ross et al. 2012)	HP-88 100 m x 0.32 mm x 0.25 µm	Initial T=75 °C increasing to 165°C at 8 °C/min, 35 min holding time, then increasing to 210 °C at 5.5 °C/min, maintain for 2 min, finally increasing to 240 °C at 15 °C/min and hold for 3 min.	FID T= 280 °C	T = 255 °C Injection volume = 1 µl	Nitrogen Flow rate = 18 ml/min
Eight retail cuts of grass-feed beef (Pavan and Duckett 2013)	SP2560 100 m x 0.25 mm x0.20 µm	Initial T=150 °C increasing to 160 °C at 1 °C/min, then increasing to 167 °C at 0.2 °C/min, after increasing to 225 °C at 1.5 °C/min and hold for 16 min.	FID T= 250 °C	T = 250 °C Injection volume = 1 µl	Hydrogen Flow rate = 1 ml/min
Intramuscular fat in organic beef (Pestana et al. 2012)	Chrompack CP-Sil 88 fused silica 100 m x 0.25 mm x0.20 µm	100 °C (held for 1 min), then increased at 50 °C/min to 150 °C (held for 20 min), then increased at 1 °C/min to 190 °C (held for 5 min) and finally increased at 1 °C/min to 200 °C (held for 35 min).	FID T= 280 °C	Split mode (1:50) T = 250 °C Injection volume = 1 µl	Helium Flow rate = 1 ml/min

6.4 Conclusion

Lipids are a diverse group of compounds that have important biological functions, namely as a source of energy and as the essential components of membranes. Humans are unable to synthesize *de novo* fatty acids of the ω-3 and ω-6 families so they must be supplied through their diet.

Meat and seafood have distinct lipid signatures. In fact, meat is characterized by high levels of SFA, namely palmitic and stearic acids, as well as by the presence of MUFA, like oleic acid. Conversely, seafood is rich in PUFA, such as DHA and EPA.

It is well established that a diet rich in SFA can lead to coronary heart diseases, type-2 diabetes and cancer while the consumption of MUFA and PUFA can serve as protection against cardiovascular morbidity and mortality. It is important that consumers are aware of the nutrient content of foods, in particular of their fat content and fatty acid profile, in order to make a conscious choice. Therefore, nutrition labeling, if applied correctly and if adequately used, can influence consumers in their purchase decisions.

Analysis of lipid extracts from food samples is an important determination in nutrition studies. This chapter focused on some of the most representative methods developed for total lipid extraction. Conventional methods for lipid extraction, such as the Folch method and the Bligh and Dyer method, among others, were revised in association with their applications to meat and seafood analysis. Also some of the most representative classical procedures for lipid analysis in meat and seafood extracts (such as the iodine value and the free fatty acid content) were discussed. Instrumental methods, such as gas chromatography and high performance liquid chromatography, were also addressed as powerful and reliable techniques for the analysis of lipids in meat and seafood extracts, mainly regarding the quantification of triacylglycerol and the fatty acid composition.

References

Abbas, K.A., Mohamed, A. and Jamilah, B. 2009. Fatty acids in fish and beef and their nutritional values: a review. Journal of Food, Agriculture & Environment. 7: 37–42.
Ackman, R.G. 1990. Seafood lipids and fatty acids. Food Reviews International. 6: 617–646.
Allen, R.R. 1955. Determination of unsaturation. J. Am. Oil Chem. Soc. 32: 671–674.
Angelo, A.J. and James Jr., C. 1993. Analysis of lipids from cooked beef by thin-layer chromatography with flame-ionization detection. J. Am. Oil Chem. Soc. 12: 1245–1250.
Antolovich, M., Prenzler, P.D., Patsalides, E., McDonald, S. and Robards, K. 2002. Methods for testing antioxidant activity. Analyst. 127: 183–198.
AOAC. 1995. Association of Official Analytical Chemists. Animal Feed. Chapter 4 in Official Methods of Analysis. 16th edition. AOAC International, Arlington, VA, USA.

192   Methods in Food Analysis

AOAC. 1995. Association of Official Analytical Chemists. Fatty acids in oils and fats, Method 969.33 Preparation of methyl esters, boron trifluoride method. 16th edition. AOAC International, Arlington, VA, USA.
AOAC. 2000. Association of Official Analytical Chemists, Method 920.160. Saponification number of oils and fats. 16th edition. AOAC International, Arlington, VA, USA.
AOAO. 2005. Official Methods of Analysis. 18th edition. Association of Official Analytical International. Washington DC, USA (cd Rom).
AOCS. 1993. Official Methods and Recommended Practices of the American Oil Chemists' Society, Method Cd 1–25. Iodine Value of Fats and Oils-Wijs Method. Champaign IL.
AOCS. 1993. Official methods and Recommended Practices of the American Oil Chemists' Society, Method Ca5a–40. Free Fatty Acids. Champaign IL.
AOCS. 1998. Official methods and Recommended Practices of the American Oil Chemists' Society, Method Cd 8–53. Peroxide value. Champaign IL.
AOCS. 2003. Official methods and Recommended Practices of the American Oil Chemists' Society, Method Cd 18–90. In: Firestone 5th edition. Champaign IL.
Arneth, W. 1998. About the determination of intramuscular fat. Fleischwirtschaft. 78: 218–220.
Aued-Pimentel, S., Kus, M.M.M., Kumagai, E.E., Ruvieri, V. and Zenebonet, O. 2010. Comparison of gas-chromatographic and gravimetric methods for quantization of total fat and fatty acids in foodstuffs. Quim. Nova 33: 76–84.
Babalola, T.O.O. and Apata, D.F. 2011. Chemical and quality evaluation of some alternative lipid sources for aqua feed production. Agric. Biol. J. N. Am. 2: 935–943.
Baily, S.K., Wells, D.E., de Boer, J. and Delbeke, K. 1994. The measurement of lipids as a co-factor for organic contaminants in biota. FRS Marine Laboratory, Aberdeen, UK.
Bandarra, N.M., Batista, I., Nunes, M.L., Empis, J.M. and Christie, W.W. 1997. Seasonal changes in lipid composition of sardine (Sardina pilchardus). Journal of Food Science. 62: 40–42.
Barriuso, B., Astiasarán, I. and Ansorena, D. 2013. A review of analytical methods measuring lipid oxidation status in foods: a challenging task. Eur. Food Res. Technol. 236: 1–15.
Batista, A., Vetter, W. and Luckas, B. 2001. Use of focused open vessel microwave-assisted extraction as prelude for the determination of the fatty acid profile of fish—a comparison with results obtained after liquid-liquid extraction according to Bligh and Dyer. Eur. Food Res. Technol. 212: 377–384.
Belkind-Gerson, J., Carreón-Rodríguez, A., Contreras-Ochoa, C., Estrada-Mondaca, S. and Parra-Cabrera, M. 2008. Fatty acids and neurodevelopment. Journal of Pediatric Gastroenterology and Nutrition. 47: S7–S9.
Berg, J.M., Tymoczko, J.L. and Stryer, L. 2007. Biochemistry. W.H. Freeman and Company, New York. USA.
Berg, H., Dahlberg, L. and Mathiasson, L. 2002. Determination of fat content and fatty acid composition in meat and meat products after supercritical fluid extraction. J. AOAC Int. 85: 1064–1069.
Berg, H., Magard, M., Johansson, G. and Mathiasson, L. 1997. Development of a supercritical fluid extraction method for determination of lipid classes and total fat in meats and its comparison with conventional methods. J. Chromatogr. 785: 345–352.
Bernárdez, M., Pastoriza, L., Sampedro, G., Herrera, J.J.R. and Cabo, M.L. 2005. Modified method for the analysis of free fatty acids in fish. J. Agric. Food Chem. 53: 1903–1906.
Bligh, E.G. and Dyer, W.J. 1959. A rapid method of total lipid extraction and purification. Can. J. Biochem. Phys. 37: 911–917.
Brondz, I. 2002. Development of fatty acid analysis by high-performance liquid chromatography, gas chromatography and related techniques. Anal. Chim. Acta. 465: 1–37.
Brum, A.A.S., de Arruda, L.F. and Regitano-d`Arce, M.A.B. 2009. Extraction methods and quality of the lipid fraction of vegetable and animal samples. Quim. Nova 32: 849–854.
BSI: Determination of total fat content in meat and meat products. BS4401-4: 1970.
Calder, P.C. 2009. Polyunsaturated fatty acids and inflammatory processes: New twists in an old tale. Biochimie. 91: 791–795.

Calder, P.C. and Yaqoob, P. 2009a. Omega-3 polyunsaturated fatty acids and human health outcomes. Biofactors. 35: 266–272.
Calder, P.C. and Yaqoob, P. 2009b. Understanding omega-3 polyunsaturated fatty acids. Postgraduate Medicine. 121: 148–157.
Campo, M.M., Nute, G.R., Hughes, S.I., Enser, M., Wood, J.D. and Richardson, R.I. 2006. Flavour perception of oxidation in beef. Meat Sci. 72: 303–311.
Carvalho, A.P. and Malcata, F.X. 2005. Preparation of fatty acid methyl esters for gas-chromatographic analysis of marine lipids: insight studies. J. Agric. Food Chem. 53: 5049–5059.
Cassileth, B. 2010. Omega-3. Oncology. (Williston Park) 24: 106.
Chardigny, J.M., Destaillats, F., Malpuech-Brugere, C., Moulin, J., Bauman, D.E., Lock, A.L., Barbano, D.M., Mensink, R.P., Bezelgues, J.B., Chaumont, P., Combe, N., Cristiani, I., Joffre, F., German, J.B., Dionisi, F., Boirie, Y. and Sebedio, J.L. 2008. Do *trans* fatty acids from industrially produced sources and from natural sources have the same effect on cardiovascular disease risk factors in healthy subjects? Results of the *trans* Fatty Acids Collaboration (TRANSFACT) study. The American Journal of Clinical Nutrition. 87: 558–566.
Christie, W.W. 1989. Gas Chromatography and Lipids: A Practical Guide. The Oily Press. Scotland.
Christie, W.W. 1993. Advances in Lipid Methodology—Two. Oily Press, Dunde, Scotland.
Christie, W.W. and Han, X. 2010. Lipid Analysis—isolation, separation, identification and lipidomic analysis. 4th ed. Oily Press, Bridgwater, U.K. and Woodhead Publishing Ltd., Cambridge, U.K.
Chukwu, O. and Imodiboh, L.I. 2009. Influence of storage conditions on shelf-life of dried beef product (kilishi). World J. Agric. Sci. 5: 34–39.
Cronin, D.A. and McKenzie, K. 1990. A rapid method for the determination of fat in foodstuffs by infrared spectrometry. Food Chem. 35: 39–49.
Darwish, D.S., van de Voort, F.R. and Smith, J.P. 1989. Proximate analysis of fish tissue by Mid-Infrared Transmission Spectroscopy. Can. J. Fish Aquatic. Sci. 46: 644–649.
Diaz, M.T., Alvarez, I., de la Fuente, J., Sanudo, C., Campo, M.M., Oliver, M.A., Font, I.F.M., Montossi, F., San Julian, R., Nute, G.R. and Caneque, V. 2005. Fatty acid composition of meat from typical lamb production systems of Spain, United Kingdom, Germany and Uruguay. Meat Sci. 71: 256–263.
de Koning, A. and Mol, T. 1991. Quantitative quality test for frozen fish: soluble protein and free fatty acid content as quality criteria for hake (*Meluccius merluccius*) stored at –18°C. J. Sci. Food Agric. 54: 449–458.
de Koning, A.J. 1999. The free fatty acid content of fish oil, part V. The effect of microbial contamination on the increase in free fatty acid content of fish oils during storage at 25°C. Eur. J. Lipid Sci. Technol. 101: 184–186.
Dobarganes, M.C. and Velasco, J. 2002. Analysis of lipid hydroperoxides. Eur. J. Lipid Sci. Technol. 104: 420–428.
Endo, Y., Tagiri-Endo, M. and Kimura, K. 2005. Rapid determination of iodine value and saponification value of fish oils by near-infrared spectroscopy. J. Food Sci. 70: C127–C131.
Enser, M., Hallett, K., Hewitt, B., Fursey, G.A. and Wood, J.D. 1996. Fatty acid content and composition of english beef, lamb and pork at retail. Meat Sci. 42: 443–456.
Estévez, M., Morcuende, D., Ventanas, S. and Cava, R. 2003. Analysis of volatiles in meat from iberian pigs and lean pigs after refrigeration and cooking by using SPME-GC-MS. J. Agric. Food Chem. 51: 3429–3435.
Eymard, S. and Genot, C. 2003. A modified xylenol orange method to evaluate formation of lipid hydroperoxides during storage and processing of small pelagic fish. Eur. J. Lipid Sci. Technol. 105: 497–501.
Ferguson, L.R. 2010. Meat and cancer. Meat Sci. 84: 308–313.

Fernandez, J., Perez-Alvarez, J. and Fernandez-Lopez, J. 1997. Thiobarbituric acid test for monitoring lipid oxidation in meat. Food Chem. 59: 345–353.
Folch, J., Less, M. and Sloane, G.H. 1957. A simple method for the isolation and purification of total lipids from animal tissues. J. Biol. Chem. 226: 497–509.
Fuchs, B., Süss, R., Teuber, K., Eibish, M. and Schiller, J. 2011. Lipid analysis by thin-layer chromatography—a review of the current state. J. Chromatograph. A. 1218: 2754–2774.
Ganhão, R., Estévez, M., Kylli, P., Heinonen, M. and Morcuende, D. 2010. Characterization of selected wild Mediterranean fruits and comparative efficacy as inhibitors of oxidative reactions in emulsified raw pork burger patties. J. Agric. Food Chem. 58: 8854–8861.
Gunnlaugsdottir, H. and Ackman, R.G. 1993. Extraction methods for determination of lipids in fish-meal—evaluation of a hexane isopropanol method as an alternative to chloroform-based methods. J. Sci. Food Agr. 61: 235–240.
Ham, B., Shelton, R., Butler, B. and Thionville, P. 1998. Calculating the iodine value for marine oils from fatty acid profiles. J. Am. Oil Chem. Soc. 75: 1445–1446.
Hara, A. and Radin, N. 1978. Lipid extraction of tissues with low-toxicity solvent. Anal. Biochem. 90: 420–426.
Harwood, J.L. and Guschina, I.A. 2009. The versatility of algae and their lipid metabolism. Biochimie. 91: 679–684.
Headlam, H.A. and Davies, M.J. 2003. Cell-mediated reduction of protein and peptide hydroperoxides to reactive free radicals. Free Radical Biology & Medicine. 34: 44–55.
Honeycutt, M.E., McFarland, V.A. and McCant, D.D. 1995. Comparison of three lipid extraction methods for fish. Bull. Environ. Contam. Toxicol. 55: 469–472.
Huang, L.C., Bulbul, U., Wen, P.C., Glew, R.H. and Ayaz, F.A. 2012. Fatty acid composition of 12 fish species from the Black Sea. J. Food Sci. 77: C512–518.
Huynh, M.D. and Kitts, D.D. 2009. Evaluating nutritional quality of pacific fish species from fatty acid signatures. Food Chem. 114: 912–918.
Indrasena, W.M., Henneberry, K., Barrow, C.J. and Kralovec, J.A. 2005. Qualitative and quantitative analysis of lipid classes in fish oils by thin-layer chromatography with an Iatroscan flame Ionization detector (TLC-FID) and liquid chromatography with an evaporative light scattering detector (LC-ELSD). Journal of Liquid Chromatography & Related Technologies. 28: 2581–2595.
Isaac, G., Waldeback, M., Eriksson, U., Odham, G. and Markides, K.E. 2005. Total lipid extraction of homogenized and intact lean fish muscles using pressurized fluid extraction and batch extraction techniques. J. Agr. Food Chem. 58: 5506–5512.
Iverson, S.J., Lang, S.L.C. and Cooper, M.H. 2001. Comparison of the Bligh and Dyer and Folch methods for total lipid determination in a broad range of marine tissue. Lipids. 36: 1283–1287.
Jakobsen, K. 1999. Dietary modifications of animal fats: status and future perspectives. Fett. (Weinheim) 101: 475–483.
James, C.S. 1995. Analytical Chemistry of Food. Blackie Academic and Professional, London. UK.
Jensen, S., Johnels, A.G., Olsson, M. and Otterlind, G. 1972. DDT and PCB in herring and cod from the Baltic, the Kattegat and the Skagerrak. Ambio. Spec. Rep. 1: 71–85.
Jensen, S., Lindqvist, D. and Asplund, L. 2009. Lipid extraction and determination of halogenated phenols and alkylphenols as their pentafluorobenzoyl derivatives in marine organisms. J. Agric. Food Chem. 57: 5872–5877.
Jensen, S., Häggberg, L., Jörundsdóttir, H. and Odham, G. 2003. A quantitative lipid extraction method for residue analysis of fish involving nonhalogenated solvents. J. Agric. Food Chem. 51: 5607–5611.
Kaitaranta, J.K. 1981. TLC-FID assessment of lipid oxidation as applied to fish lipids rich in triglycerides. J. Am. Oil Chem. Soc. 58: 710–713.
Kanatt, S.R., Paul, P., D'Souza, S.F. and Thomas, P. 1998. Lipid peroxidation in chicken meat during chilled storage as affected by antioxidants combined with low-dose gamma irradiation. J. Food Sci. 63: 198–200.

Karl, H., Oehlenschaeger, J., Bekaert, K., Berge, J.P., Cadun, A., Duflos, G., Poli, B.M., Tejada, M., Testi, S. and Timm-Heinrich, M. 2012. WEFTA Interlaboratory comparison on total lipid determination in fishery products using the Smedes method. J. AOAC Int. 95: 489–493.

Kataoka, H., Lord, H.L. and Pawliszyn, J. 2000. Applications of solid-phase microextraction in food analysis. J. Chromatogr. A. 880: 35–62.

Ke, P.J. and Woyewoda, A.D. 1978. A titrimetric method for determination of free fatty acids in tissues and lipids with ternary solvents and m-cresol purple indicator. Anal. Chim. Acta. 99: 387–391.

King, J.W. 2002. Supercritical Fluid Extraction: Present Status and Prospects. Grasas Aceites. 53: 8–21.

Knothe, G. 2002. Structure indices in FA chemistry. How relevant is the iodine value? J. Am. Oil Chem. Soc. 79: 847–854.

Kondo, T., Yamamoto, H., Tatarazako, N., Kawabe, K., Koshio, M., Hirai, N. and Morita, M. 2005. Bioconcentration factor of relatively low concentrations of chlorophenols in Japanese medaka. Chemosphere. 61: 1299–1304.

Kouba, M. and Mourot, J. 2011. A review of nutritional effects on fat composition of animal products with special emphasis on n-3 polyunsaturated fatty acids. Biochimie. 93: 13–17.

Ladikos, D. and Lougovois, V. 1990. Lipid oxidation in muscle foods: a review. Food Chem. 35: 295–314.

Lee, H., Kizito, S.A., Weese, S.J., Craig-Schmidt, M.C., Lee, Y., Wei, C.-I. and An, H. 2003. Analysis of headspace volatile and oxidized volatile compounds in DHA-enriched fish oil on accelerated oxidative storage. J. Food Sci. 68: 2169–2177.

Li, Y. and Watkins, B.A. 2001. Analysis of fatty acids in food lipids. Current Protocols in Food Analytical Chemistry.

Lichtenstein, A.H., Kennedy, E., Barrier, P., Danford, D., Ernst, N.D., Grundy, S.M., Leveille, G.A., Van Horn, L., Williams, C.L. and Booth, S.L. 1998. Dietary fat consumption and health. Nutr. Rev. 56: S3–19; discussion S19–28.

Liu, K.-S. 1994. Preparation of fatty acid methyl esters for gas-chromatographic analysis of lipids in biological materials. J. Am. Oil Chem. Soc. 71: 1179–1187.

Lowry, R.R. and Tinsley, I.J. 1976. Rapid colorimetric determination of free fatty acids. J. Am. Oil Chem. Soc. 53: 470–472.

Lunn, J. and Theobald, H.E. 2006. The health effects of dietary unsaturated fatty acids. Nutrition Bulletin. 31: 178–224.

Ma, Y., Wang, L., Sun, Y.M., Yu, Y.H. and Zhu, B.W. 2006. Research on extracting lipid out of scallop viscera by supercritical fluid extraction. Food and Fermentation Industries. 32: 156–159.

MacArtain, P., Gill, C.I.R., Brooks, M., Campbell, R. and Rowland, I.R. 2007. Nutritional value of edible seaweeds. Nutrition Reviews. 65: 535–543.

Maid-Kohnert, U. 2002. Lexikon der Ernährung. Spektrum Akademischer Verlag, Heidelberg.

Manirakiza, P., Covaci, A. and Schepens, P. 2001. Comparative study on total lipid determination using Soxhlet, Roese-Gottlieb, Bligh & Dyer, and Modified Bligh & Dyer extraction methods. J. Food Compod. Anal. 14: 93–100.

Mathias, J.A., Williams, P.C. and Sobering, D.C. 1987. The determination of lipid and protein in freshwater fish using near-infrared reflectance spectroscopy. Aquaculture. 61: 303–311.

McAfee, A.J., McSorley, E.M., Cuskelly, G.J., Moss, B.W., Wallace, J.M.W., Bonham, M.P. and Fearon, A.M. 2010. Red meat consumption: An overview of the risks and benefits. Meat Sci. 84: 1–13.

McLean, B. and Drake, P. 2002. Review of methods for the determination of fat and oil in foodstuffs. Review No. 37. Campden & Chorleywood Food Research Association Group (CCFRA).

Meat Research Corporation 1998. Iodine value. Australian Meat Technology.

Meyer, B.J. 2011. Are we consuming enough long chain omega-3 polyunsaturated fatty acids for optimal health? Prostaglandins, Leukotrienes, and Essential Fatty Acids. 85: 275–280.
Min, B. and Ahn, D.U. 2005. Mechanism of lipid peroxidation in meat and meat products—a review. Food Sci. Biotechnol. 14: 152–163.
Miyasaki, T., Hamaguchi, M. and Yokoyama, S. 2011. Change of volatile compounds in fresh fish meat during ice storage. J. Food Sci. 76: C1319–C1325.
Morrison, W.R. and Smith, L.M. 1964. Preparation of fatty acid methyl esters and dimethylacetals from lipids with boron fluoride methanol. J. Lipid Res. 5: 600–608.
Motard-Belanger, A., Charest, A., Grenier, G., Paquin, P., Chouinard, Y., Lemieux, S., Couture, P. and Lamarche, B. 2008. Study of the effect of trans fatty acids from ruminants on blood lipids and other risk factors for cardiovascular disease. The American Journal of Clinical Nutrition. 87: 593–599.
Mozaffarian, D. and Wu, J.H. 2012. (n-3) fatty acids and cardiovascular health: are effects of EPA and DHA shared or complementary? The Journal of Nutrition. 142: 614S–625S.
Murty, V.R., Bhat, J. and Muniswaran, P.K.A. 2002. Hydrolysis of oils by using immobilized lipase enzyme: a review. Biotechnol. Bioprocess Eng. 7: 57–66.
Nanton, D.A., Vegusdal, A., Rørå, A.M.B., Ruyter, B., Baeverfjord, G. and Torstensen, B.E. 2007. Muscle lipid storage pattern, composition, and adipocyte distribution in different parts of Atlantic salmon (*Salmo salar*) fed fish oil and vegetable oil. Aquaculture. 265: 230–243.
Nelson, D.L. and Cox, M.M. 2005. Lehninger—Principles of Biochemistry. W.H. Freeman and Company, New York. USA.
Nevigato, T., Masci, M., Orban, E., Di Lena, G., Casini, I. and Caproni, R. 2012. Analysis of fatty acids in 12 mediterranean fish species: advantages and limitations of a new GC-FID/GC–MS based technique. Lipids. 47: 741–753.
Niki, E. and Yoshida, Y. 2005. Biomarkers for oxidative stress: measurement, validation, and application. The Journal of Medical Investigation: JMI. 52 Suppl.: 228–230.
Niki, E., Yoshida, Y., Saito, Y. and Noguchi, N. 2005. Lipid peroxidation: mechanisms, inhibition, and biological effects. Biochemical and Biophysical Research Communications. 338: 668–676.
Orban, E., Di Lena, G., Ricelli, A., Paoletti, F., Casini, I., Gambelli, L. and Caproni, R. 2000. Quality characteristics of sharpsnout sea bream (*Diplodus puntazzo*) from different intensive rearing systems. Food Chem. 70: 27–32.
Osman, H., Suriah, A.R. and Law, E.C. 2001. Fatty acid composition and cholesterol content of selected marine fish in Malaysian waters. Food Chem. 73: 55–60.
Ospina, E.J., Sierra, C.A., Ochoa, O., Perez-Alvarez, J.A. and Fernandez-Lopez, J. 2012. Substitution of saturated fat in processed meat products: a review. Critical Reviews in Food Science and Nutrition. 52: 113–122.
Ould Ahmed Louly, A.W., Gaydou, E.M. and Ould El Kebir, M.V. 2011. Muscle lipids and fatty acid profiles of three edible fish from the Mauritanian coast: *Epinephelus aeneus*, *Cephalopholis taeniops* and *Serranus scriba*. Food Chem. 124: 24–28.
Ozogul, Y., Simsek, A., Balikci, E. and Kenar, M. 2012. The effects of extraction methods on the contents of fatty acids, especially EPA and DHA in marine lipids. Int. J. Food Sci. Nutr. 63: 326–331.
Ozogul, Y., Ozogul, F., Cicek, E., Polat, A. and Kuley, E. 2009. Fat content and fatty acid compositions of 34 marine water fish species from the Mediterranean Sea. International Journal of Food Sciences and Nutrition. 60: 464–475.
Papastergiadis, A., Mubiru, E., Van Langenhove, H. and De Meulenaer, B. 2012. Malondialdehyde measurement in oxidized foods: evaluation of the spectrophotometric thiobarbituric acid reactive substances (TBARS) test in various foods. J. Agric. Food Chem. 60: 9589–9594.
Pauwels, E.K., Volterrani, D., Mariani, G. and Kairemo, K. 2009. Fatty acid facts, Part IV: docosahexaenoic acid and Alzheimer's disease. A story of mice, men and fish. Drug News & Perspectives. 22: 205–213.

Pavan, E. and Duckett, S.K. 2013. Fatty acid composition and interrelationships among eight retail cuts of grass-feed beef. Meat Sci. 93: 371–377.
Perez-Palacios, T., Ruiz, J., Martin, D., Muriel, E. and Antequera, T. 2008. Comparison of different methods for total lipid quantification in meat and meat products. Food Chem. 110: 1025–1029.
Perez-Palacios, T., Ruiz, J., Ferreira, I.M.P.L.V.O., Petisca, C. and Antequera, T. 2012. Effect of solvent to sample ratio on total lipid extracted and fatty acid composition in meat products within different fat content. Meat Sci. 91: 369–373.
Pestana, J.M., Costa, A.S., Alves, S.P., Martins, S.V., Alfaia, C.M., Bessa, R.J. and Prates, J.A. 2012. Seasonal changes and muscle type effect on the nutritional quality of intramuscular fat in Mirandesa-PDO veal. Meat Sci. 90: 819–827.
Pestana, J.M., Costa, A.S.H., Martins, S.V., Alfaia, C.M., Alves, S.P., Lopes, P.A., Bessa, R.J.B. and Prates, J.A.M. 2012. Effect of slaughter season and muscle type on the fatty acid composition, including conjugated linoleic acid isomers, and nutritional value of intramuscular fat in organic beef. J. Sci. Food. Agric. 92: 2428–2435.
Pickova, J. 2009. Importance of knowledge on lipid composition of foods to support development towards consumption of higher levels of n-3 fatty acids via freshwater fish. Physiological Research/Academia Scientiarum Bohemoslovaca. 58 Suppl. 1: S39–45.
Pillonel, L., Bosset, J.O. and Tabacchi, R. 2002. Rapid preconcentration and enrichment techniques for the analysis of food volatile. A review. Food Science and Technology. 35: 1–14.
Pocklington, W.D. 1990. Determination of the iodine value of oils and fats. Results of a collaborative study. Pure & Appl. Chem. 62: 2339–2343.
Pomeranz, Y. and Meloan, C.E. 1994. Food analysis: theory and practice. Chapman & Hall, New York.
Prato, E. and Biandolino, F. 2012. Total lipid content and fatty acid composition of commercially important fish species from the Mediterranean, Mar Grande Sea. Food Chem. 131: 1233–1239.
Pravst, I., Žmitek, K. and Žmitek, J. 2010. Coenzyme Q10 contents in foods and fortification strategies. Critical Reviews in Food Science and Nutrition. 50: 269–280.
Prevolnik, M., Candek-Potokar, M., Skorjanc, D., Velikonja-Bolta, S., Skrlep, M., Znidarsic, T. and Babnik, D. 2005. Predicting intramuscular fat content in pork and beef by near infrared spectroscopy. J. Near Infrared Spec. 13: 77–85.
Raes, K. and De Smet, S. 2009. Fatty acids. pp. 141–154. In: Nollet, L.M.L. and Toldrá, F. (eds.). Handbook of Muscle Food Analysis. CRC Press, Boca Raton, USA.
Ramalhosa, M.J., Paíga, P., Morais, S., Alves, M.R., Delerue-Matos, C. and Oliveira, M.B.P.P. 2012. Lipid content of frozen fish: Comparison of different extraction methods and variability during freezing storage. Food Chem. 131: 328–336.
Remig, V., Franklin, B., Margolis, S., Kostas, G., Nece, T. and Street, J.C. 2010. *Trans* fats in America: A review of their use, consumption, health implications, and regulation. Journal of the American Dietetic Association. 110: 585–592.
Rinchard, J., Czesny, S. and Dabrowski, K. 2007. Influence of lipid class and fatty acid deficiency on survival, growth, and fatty acid composition in rainbow trout juveniles. Aquaculture. 264(1–4): 363–371.
Rohman, A., Triyana, K., Sismindari and Erwanto, Y. 2012. Differentiation of lard and other animal fats based on triacylglycerols composition and principal component analysis. International Food Research Journal. 19: 475–479.
Róldan, H., Barassi, C. and Trucco, R. 1985. Increase on free fatty acids during ripening of anchovies (*Engraulis anchoita*). J. Food Technol. 20: 581–585.
Ross, G.R., Van Nieuwenhove, C.P. and González, S.N. 2012. Fatty acid profile of pig meat after probiotic administration. J. Agric. Food Chem. 60: 5974–5978.
Rubio-Rodriguez, N., de Diego, S.M., Beltran, S., Jaime, I., Sanz, M.T. and Rovira, J. 2012. Supercritical fluid extraction of fish oil from fish by-products: A comparison with other extraction methods. J. Food Eng. 109: 238–248.

Ruíz, A., Cañada, M.J.A. and Lendl, B. 2001. A rapid method for peroxide value determination in edible oils based on flow analysis with Fourier transform infrared spectroscopic detection. Analyst. 126: 242–246.
Sahena, F., Zaidul, I.S.M., Jinap, S., Yazid, A.M., Khatib, A. and Norulaini, N.A. 2010. Fatty acid compositions of fish oil extracted from different parts of Indian mackerel (*Rastrelliger kanagurta*) using various techniques of supercritical CO_2 extraction. Food Chem. 120: 879–885.
Salih, A., Smith, D., Price, J. and Dawson, L. 1987. Modified extraction 2-thiobarbituric acid method for measuring lipid oxidation in poultry. Poult. Sci. 66: 1483–1488.
Sarker, M.Z.I., Selamat, J., Habib, A.S.M.A., Ferdosh, S., Akanda, M.J.H. and Jaffri, J.M. 2012. Optimization of supercritical CO_2 extraction of fish oil from viscera of African catfish (*Clarias gariepinus*). Int. J. Mol. Sci. 13: 11312–11322.
Schlechtriem, C., Focken, U. and Becker, K. 2003. Effect of different lipid extraction methods on delta C-13 of lipid and lipid-free fractions of fish and different fish feeds. Isot. Environ. Health S. 39: 135–140.
Schlechtriem, C., Henderson, R.J. and Tocher, D.R. 2008. A critical assessment of different transmethylation procedures commonly employed in the fatty acid analysis of aquatic organisms. Limnol. Oceanogr. Methods 6: 523–531.
Schlechtriem, C., Fliedner, A. and Schäfers, C. 2012. Determination of lipid content in fish samples from bioaccumulation studies: contributions to the revision of guideline OECD 305. Environmental Sciences Europe. 24: 13.
Schmid, A. 2010. The role of meat fat in the human diet. Critical Reviews in Food Science and Nutrition. 51: 50–66.
Schuchardt, U. and Lopes, O.C. 1988. Tetramethylguanidine catalysed transesterification of fats and oils: a new method for rapid determination of their composition. J Am. Oil Chem. Soc. 65: 1940–1941.
Schwalme, K., Mackay, W.C. and Clandinin, M.T. 1993. Seasonal dynamics of fatty acid composition in female northern pike (*Esox lucius* L.). J. Comp. Physiol. B. 163: 277–287.
Shahidi, F. and Wanasundara, P.K.J.P.D. 2008. Extraction and analysis of lipids. pp. 125–149. In: Akoh, C.C. and Min, D.B. (eds.). Food Lipids Chemistry, Nutrition, and Biotechnology. CRC Press, Boca Raton, USA.
Shahidi, F. and Wanasundara, P.K.J.P.D. 2008. Methods for measuring oxidative rancidity in fats and oils. pp. 387–407. In: Akoh, C.C. and Min, D.B. (eds.). Food Lipids Chemistry, Nutrition, and Biotechnology. CRC Press, Boca Raton, USA.
Shahidi, F. and Zhong, Y. 2005. Lipid oxidation: measurement methods. pp. 357–385. In: Shahidi, F. (ed.). Bailey's Industrial Oil and Fat Products. John Wiley & Sons, Inc., New Jersey.
Shantha, N.C. 1992. Thin-layer chromatography-flame ionization detection Iatroscan system. J. Chromatography 624: 21–35.
Shantha, N.C. and Decker, E.A. 1994. Rapid, sensitive, iron-based spectrophotometric methods for determination of peroxide values of food lipids. Journal of AOAC International. 77: 421–424.
Shantha, N.C. and Napolitano, G.E. 1992. Gas chromatography of fatty acids. J. Chromatogr. 624: 37–51.
Shantha, N.C. and Napolitano, G.E. 1998. Lipid analysis using thin-layer chromatography and the Iatroscan. pp. 1–33. In: Hamilton, R.J. (ed.). Lipid Analysis of Oils and Fats. Chapman & Hall, UK.
Sheppard, A.J. 1963. Suitability of lipid extraction procedures for gas-liquid chromatography. J. Am. Oil Chem. Soc. 40: 545–548.
Sherazi, S.T.H., Mahesar, S.A. and Bhanger, M.I. 2007. Rapid determination of free fatty acids in poultry feed lipid extracts by SB-ATR FTIR spectroscopy. J. Agric. Food Chem. 55: 4928–4932.
Sinanoglou, V.J., Strati, I.F., Bratakos, S.M., Proestos, C., Zoumpoulakis, P. and Miniadis-Meimaroglou, S. 2013. On the combined application of Iatroscan TLC-FID and GC-FID

to identify total, neutral, and polar lipids and their fatty acids extracted from foods. ISRN Chromatography. 2013: 1–8.
Sirot, V., Oseredczuk, M., Bemrah-Aouachria, N., Volatier, J.L. and Leblanc, J.C. 2008. Lipid and fatty acid composition of fish and seafood consumed in France: CALIPSO study. Journal of Food Composition and Analysis. 21: 8–16.
Siriwardhana, N., Kalupahana, N.S. and Moustaid-Moussa, N. 2012. Chapter 13 - Health Benefits of n-3 Polyunsaturated Fatty Acids: Eicosapentaenoic Acid and Docosahexaenoic Acid. In: Se-Kwon K., editor. Advances in Food and Nutrition Research: Academic Press. 65: 211–22.
Smedes, F. 1999. Determination of total lipid using non-chlorinated solvents. Analyst. 124: 1711–1718.
Smedes, F. and Askland, T.K. 1999. Revisiting the development of the Bligh and Dyer total lipid determination method. Mar. Pollut. Bull. 38: 193–201.
Soriguer, F., Serna, S., Valverde, E., Hernando, J., Martin-Reyes, A., Soriguer, M., Pareja, A., Tinahones, F. and Esteva, I. 1997. Lipid, protein, and calorie content of different Atlantic and Mediterranean fish, shellfish, and molluscs commonly eaten in the south of Spain. European Journal of Epidemiology. 13: 451–463.
Soxhlet, F. 1879. Die gewichtsanalytische Bestimmung des Milchfettes. Polytechnisches J. (Dingler's) 232–461.
Strobel, C., Jahreis, G. and Kuhnt, K. 2012. Survey of n-3 and n-6 polyunsaturated fatty acids in fish and fish products. Lipids in Health and Disease. 11: 144.
Sun, Y.E., Wang, W.D., Chen, H.W. and Li, C. 2011. Autoxidation of unsaturated lipids in food emulsion. Critical Reviews in Food Science and Nutrition. 51: 453–466.
Tanamati, A., Oliveira, C.C., Visentainer, J.V., Matsushita, M. and de Souza, N.E. 2005. Comparative study of total lipids in beef using chlorinated solvent and low-toxicity solvent Methods. J. Am. Oil Chem. Soc. 82: 393–397.
Tocher, D.R., Zheng, X., Schlechtriem, C., Hastings, N., Dick, J.R. and Teale, A.J. 2006. Highly unsaturated fatty acid synthesis in marine fish: cloning, functional characterization, and nutritional regulation of fatty acyl delta 6 desaturase of Atlantic cod (*Gadus morhua* L.). Lipids. 41: 1003–1016.
Tranchida, P.Q., Donato, P., Dugo, P., Dugo, G. and Mondell, L. 2007. Comprehensive chromatographic methods for the analysis of lipids. Trends in Analytical Chemistry. 26: 191–205.
Tufan, B., Koral, S. and Köse, S. 2011. Changes during fishing season in the fat content and fatty acid profile of edible muscle, liver and gonads of anchovy (*Engraulis encrasicolus*) caught in the Turkish Black Sea. International Journal of Food Science & Technology. 46: 800–810.
Valsta, L.M., Tapanainen, H. and Mannisto, S. 2005. Meat fats in nutrition. Meat Sci. 70: 525–530.
van Ginneken, V.J., Helsper, J.P., de Visser, W., van Keulen, H. and Brandenburg, W.A. 2011. Polyunsaturated fatty acids in various macroalgal species from North Atlantic and tropical seas. Lipids in Health and Disease. 10: 104.
van Poppel, G., van Erp-Baart, M.A., Leth, T., Gevers, E., Van Amelsvoort, J., Lanzmann-Petithory, D., Kafatos, A. and Aro, A. 1998. *Trans* fatty acids in foods in Europe: The TRANSFAIR Study. Journal of Food Composition and Analysis. 11: 112–136.
Vinci, G., Tantini, C., Vezzio, D., Rossi, M. and D`Ascenzo, F. 1999. Protein and fat determination in meat and meat-based products by means of rapid and precise analytical methods. Industrie Alimentari. 38: 369–373.
Wall, R., Ross, R.P., Fitzgerald, G.F. and Stanton, C. 2010. Fatty acids from fish: the anti-inflammatory potential of long-chain omega-3 fatty acids. Nutrition Reviews. 68: 280–289.
Wilkes, J.G., Conte, E.D., Kim, Y., Holcomb, M., Sutherland, J.B. and Miller, D.W. 2000. Sample preparation for the analysis of flavors and off-flavors in foods. J. Chromatogr. A. 880: 3–33.

Wrolstad, R.E., Acree, T.E., Decker, E.A., Penner, M.H., Reid, D.S., Schwartz, S.J., Shoemaker, C.F., Smith, D. and Sporns, P. 2005. Handbook of food analytical chemistry: water, proteins, enzymes, lipids and carbohydrates. John Wiley & Sons Inc., New Jersey.

Wu, J.P. and Kohler, P. 2000. Estimation of fatty acid profiles in musculus longissimus dorsi of sheep—Comparative study of two fat extraction methods. Fleischwirtschaft. 80: 123–125.

Xiao, L.P., Mjos, S.A. and Haugsgjerd, B.O. 2012. Efficiencies of three common lipid extraction methods evaluated by calculating mass balances of the fatty acids. J. Food Compos. Anal. 25: 198–207.

Zhou, D.Y., Zhu, B.W., Tong, L., Wu, H.T., Qin, L., Tan, H., Chi, Y.L., Qu, J.Y. and Murata, Y. 2010. Extraction of lipid from scallop (*Patinopecten yessoensis*) viscera by enzyme-assisted solvent and supercritical carbon dioxide methods. Int. J. Food Sci. Tech. 45: 1787–1793.

Zlatanos, S. and Laskaridis, K. 2007. Seasonal variation in the fatty acid composition of three Mediterranean fish—sardine (*Sardina pilchardus*), anchovy (*Engraulis encrasicholus*) and picarel (*Spicara smaris*). Food Chem. 103: 725–728.

7

Vibrational and Electronic Spectroscopy and Chemometrics in Analysis of Edible Oils

Ewa Sikorska,[1,*] *Igor Khmelinskii*[2] *and Marek Sikorski*[3]

ABSTRACT

This chapter presents applications of vibrational and electronic spectroscopic techniques coupled with chemometrics in studies of edible oils. The spectra of edible oils are characterized–including absorption in ultraviolet (UV), visible (Vis) and near (NIR) and mid (MIR) infrared regions, fluorescence in UV-Vis, and Raman scattering spectra. Qualitative and quantitative multivariate methods of spectral analysis are presented. Chemometric analysis of the spectral data enables establishment of calibration and classification and/or discrimination models that allow prediction of a variety of oil properties based on the spectral measurements.

---

[1] Faculty of Commodity Science, Poznań University of Economics, Poland.
[2] University of Algarve, FCT, DQF and CIQA, Portugal.
[3] Faculty of Chemistry, Adam Mickiewicz University, Poznań, Poland.
* Corresponding author

7.1 Introduction

Spectroscopic techniques may be an efficient alternative to the classical chemical methods used for oil analysis. Easiness and simplicity of measurements are the most attractive features of spectroscopy from the practical point of view. The application of spectroscopy enables development of analytical procedures that are rapid and clean, avoiding extensive sample pretreatment and usage of reagents.

The main drawback of spectroscopy is spectral overlapping of signals of the analyte(s) of interest with those of other components of the sample matrix. Therefore, multivariate methods have to be used, such as multivariate calibration or classification, in order to extract analytically useful information from spectra. This chapter presents analytical applications of spectroscopic techniques coupled with multivariate analysis in studies of edible oils. Vibrational and electronic spectra of edible oils are characterized, including absorption in ultraviolet (UV), visible (Vis) and near (NIR)- and mid (MIR)-infrared regions, fluorescence in UV-Vis, and Raman scattering spectra. Multivariate methods of spectral analysis are presented. Practical applications of spectroscopy and chemometrics in studies of various aspects of quality of edible oils are reviewed.

7.2 Spectral Characteristics of Edible Oils

7.2.1 Overview of Spectroscopic Techniques

Spectroscopy is based on the interaction of the electromagnetic radiation and matter. The way the radiation interacts with molecules provides information about molecular properties and is utilized for basic research and analytical purposes.

Molecular spectroscopy comprises a diversity of techniques, see Table 7.1. Depending on the type of interaction between light and matter, spectroscopic techniques are divided into absorption, emission and scattering spectroscopy. The energy of atoms and molecules is quantized. A photon with the energy matching the difference between particular energy levels in a molecule is absorbed. As a result, the molecule reaches a higher (excited) energy level. The excited molecule may deactivate in various ways, one of them is emission of a photon with a different energy. When the photon energy is different from energy differences between molecular levels, it is not absorbed, with some of incident photons scattered.

Another classification of spectroscopic techniques is based on the energy of electromagnetic radiation that interacts with molecules and on the changes in molecular energy induced by these interactions. Depending on the energy of electromagnetic radiation, different transitions may occur.

Table 7.1 Characteristics of molecular spectroscopic techniques in ultraviolet, visible and infrared region.

Wavelength range	Wavelength of radiation	Interaction with matter	Types of energy transition
Ultraviolet (UV)	190–380 nm	Absorption and emission	Bonding electrons in molecules
Visible (Vis)	380–800 nm	Absorption and emission	Bonding electrons in molecules
Infrared (IR)	0.80–1000 μm	Absorption, Raman scattering	Vibrations of atoms in molecular bonds

The radiation in UV and Vis ranges induces electronic transitions, changing the distribution of binding electrons in molecules. The radiation in NIR and MIR ranges induces vibrational transitions.

The results of interaction of molecules with light are assessed by measuring the spectra. The characteristics of spectra, including the number of bands, their location, intensities, and shapes are "unique" for a particular molecule, and are used to obtain qualitative and quantitative information.

Multivariate methods are used to extract quantitative, qualitative, or structural information from spectra, including data reduction, classification and regression techniques.

7.2.2 Vibrational Spectroscopy of Oils

The infrared range of electromagnetic radiation (0.8–1000 μm) is divided into near-IR (NIR, 0.8–2.5 μm), mid-IR (MIR, 2.5–25 μm) and far-IR (FIR, 25–1000 μm). NIR and IR ranges are most frequently used in food analysis (Nielsen 2010). The photon energy of infrared radiation corresponds to differences between vibrational levels in molecules. Absorption of a MIR photon typically excites one of the fundamental vibrations, associated with a change of the dipole moment of an oscillating molecule. Note that the molecule passes from the fundamental to the first excited vibrational state, within the electronic ground state of a molecule. The MIR spectra comprise well-defined, intense bands which correspond to the fundamental vibrations of specific functional groups. The first infrared spectra of fatty acids and vegetable oils were published by Coblentz in 1905 (McClure 2003). Since then, infrared spectroscopy has been widely used in lipid research and as a routine tool in the fats and oils industry.

The absorption of energy in the NIR region corresponds to the overtones and combinations of the vibrational modes involving C–H, O–H, and N–H chemical bonds.

When the energy of radiation does not match the energy levels in a molecule, scattering may occur instead of absorption. The process may be elastic (Rayleigh scattering), when scattered radiation has the same energy as incident radiation, or inelastic (Raman scattering). In Raman scattering the molecules that are excited from the fundamental vibrational state to the virtual unstable state upon interaction with photon, return to a higher (excited) vibrational state. The difference in energy between incident and scattered radiation corresponds to one of the frequencies of molecular vibrations. Active transitions involve changes in polarizability of molecules and do not require changes in dipole moment. Therefore, Raman spectroscopy is complementary to MIR; some vibrations are active only in Raman or MIR spectroscopy, while others are observed in both. For example, symmetrical vibrations not active in MIR are active in Raman spectroscopy. Raman spectroscopy usually probes for energy differences in the 4000–100 cm^{-1} range.

Raman spectra are measured by illuminating the sample with laser radiation within the near-infrared, visible or ultraviolet range, and monitoring the radiation scattered by the sample. The intensity of the scattered radiation is plotted as a function of difference between energy of scattered and incident light ($\Delta \tilde{v}$ in cm^{-1}). Because the Raman effect is typically very weak, different techniques are used to enhance the measured signal, including resonance Raman (RR) spectroscopy, surface-enhanced Raman scattering (SERS), and tip-enhanced Raman scattering (TERS) (Ellis et al. 2012). Today more than 25 different methods involving Raman spectroscopy are known (Long 2002).

Mid-infrared Absorption Spectra of Oils

MIR spectrum (Fig. 7.1) of a vegetable oil contains well-resolved intense bands that correspond to the fundamental vibrations of specific functional groups of the main oil constituents (Table 7.2).

Only one band of the first overtone of the C=O stretching vibration of triacyloglycerol ester bond is observed in fresh, pure and dry oil in the 4000–3050 cm^{-1} zone. The bands of the C-H stretching vibration are present in the 3000–2800 cm^{-1} range. The very intense bands with their maxima at 2924 and 2853 cm^{-1} are ascribed respectively to asymmetric and symmetric stretching vibrations of CH$_2$ groups of the fatty acid chains in triacyloglycerols (Guillén and Cabo 1997). The stretching vibrations of CH$_3$ groups appears as the shoulder in this region at 2962 to 2872 cm^{-1} (Vlachos et al. 2006).

The bands corresponding to the CH stretching absorption of *cis* double bonds (CH=CH) at 3006 cm^{-1} are observed at the high-energy end of the 3000–2800 cm^{-1} range. The absorption band of *trans* double bonds occurs at a slightly higher frequency, 3025 cm^{-1}, but is very weak (Guillén and Cabo 1997).

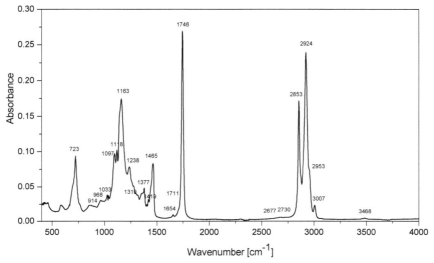

Figure 7.1 Infrared absorption spectrum of olive oil.

An intense band at 1746 cm^{-1} characteristic for oils and fats corresponds to the C=O stretching vibrations of the carbonyl group in ester linkages between fatty acids and glycerol backbone. A band arising from the carboxyl groups (RCOOH) of the free fatty acids appears as a shoulder at the low-frequency side of this band. The 1400–1200 cm^{-1} range comprises the bands associated with bending vibrations of CH$_2$ and CH$_3$ groups. The characteristic bands of the stretching vibrations of C-O and C-C groups are present in the 1125–1095 cm^{-1} range.

A band of OH stretching vibrations appears in the spectra of oils undergoing auto-oxidation in presence of moisture, hydroperoxides (ROOH), or their breakdown product alcohols in the 3700–3400 cm^{-1} range. The secondary lipid oxidation products, aldehydes (RCHO) and ketones (RCOR′) have weak absorption just beyond the lower end of the 3025–2850 cm^{-1} range. Other absorption bands due to carbonyl aldehydes or ketones may appear in the 1730–1650 cm^{-1} range (Van de Voort et al. 2001).

In the fingerprint region, between 1500 and 900 cm^{-1}, several absorption bands appear. The band at 1460 cm^{-1} arises from CH$_2$ and CH$_3$ scissoring vibrations, at 1378 cm^{-1}—from CH$_3$ bending vibration. The bands at 1238, 1163, 1118, and 1097 cm^{-1} originate from C-O stretching vibrations. At 968 cm^{-1} the band of isolated *trans* double bonds is present, assigned to the C-H out-of-plane deformation (Guillén and Cabo 1997). It is highly characteristic and is utilized by official methods for determination of total isolated *trans* fatty acids in fats and oils (Commission Regulation (EU) 2011).

Table 7.2 Functional groups and modes of vibrations in oils MIR spectra (Guillén and Cabo 1997; Lerma-García et al. 2011).

Theoretical cm^{-1}	Functional group	Mode of vibration
3468	-C=O (ester)	Overtone
3025s	=C-H *trans-*	Stretching
3006	=C-H *cis-*	Stretching
2953s	-C-H (CH$_3$)	Stretching asym
2924	-C-H (CH$_2$)	Stretching asym
2853	-C-H (CH$_2$)	Stretching sym
2730	-C=O ester	Fermi resonance
2677	-C=O ester	Fermi resonance
1746	-C=O ester	Stretching
1711s	-C=O acid	Stretching
1654	-C=C- *cis*	Stretching
1648	-C=C- *cis*	Stretching
1465	-C-H (CH$_2$, CH$_3$)	Bending (scissoring)
1418	=C-H *cis-*	Bending (rocking)
1400		Bending
1377	-C-H (CH$_3$)	Bending sym
1359	O-H	Bending in plane
1319		Bending
1238	-C-O, -CH$_2$-	Stretching, bending
1163	-C-O, -CH$_2$-	Stretching, bending
1138	-C-O	Stretching
1118	-C-O	Stretching
1097	-C-O	Stretching
1033s	-C-O	Stretching
968	-HC=CH- *trans-*	Bending out of plane
914	-HC=CH- *cis-*	Bending out of plane
850	=CH$_2$	Wagging
723	-(CH$_2$)$_n$-, -HC=CH- (*cis*)	Bending rocking

s-shoulder

The corresponding bands of conjugated *trans/trans* and *cis/trans* dienes are shifted to slightly higher frequencies. An intense band observed at 723 cm^{-1} is assigned to the CH$_2$ rocking vibration. On the low frequency side appears the C–H out-of-plane deformation band of *cis* double bonds (Van de Voort et al. 2001).

Near-infrared Absorption Spectra of Oils

NIR spectra of oils (Fig. 7.2, Table 7.3) in the range of 800–2500 nm contain broad overlapping bands corresponding mainly to overtones and combinations of the vibrational modes involving C–H, O–H, and N–H chemical bonds, which correspond to the fundamental vibrations observed in MIR. The NIR spectroscopy provides information about chemical structure; however, coupled with chemometrics, it is widely used for analysis of oils. For example, the method for the determination of the iodine value (IV) using NIR spectroscopy was accepted by the American Oil Chemists' Society (AOCS) in 2000 (Van de Voort et al. 2001).

The combination bands of the C–H stretching vibration with other vibrational modes occur in the 4000–4500 cm^{-1} range (Pereira et al. 2008). A band with the two maxima at 5787 and 5671 cm^{-1} corresponds to the first overtones of the C-H stretching vibrations in methyl, methylene and ethylene groups (Sinelli et al. 2010a). Oleic acid absorbs at 5797 cm^{-1} and saturated and *trans*-unsaturated triacylglycerols absorb at 5797 and 5681 cm^{-1} (Sinelli et al. 2010b). The bands corresponding to the second overtone of the C=O stretching vibrations occur at 5260 and 5178 cm^{-1} (Woodcock et al. 2008). The low-intensity bands at 4659 and 4572 cm^{-1} are assigned to the combination of C-H and C=O stretching vibrations (Sinelli et al. 2010a) and at 4662 and 4596 cm^{-1}—are associated with the presence of -CH=CH- double bonds (Woodcock et al. 2008).

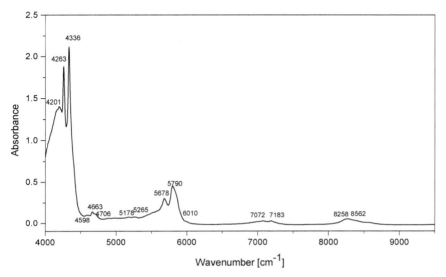

Figure 7.2 Near-infrared absorption spectrum of olive oil.

Table 7.3 Functional groups and modes of vibrations in oils NIR spectra (Woodcock et al. 2008).

Theoretical cm^{-1}	Theoretical nm	Functional group	Mode of vibration
8562	1168	CH$_3$–	C-H stretching 2nd overtone
8258	1211	–CH$_2$–	C-H stretching 2nd overtone
7189	1391	CH$_3$–	2C-H stretching and C-H deformation
7072	1414	–CH$_2$–	Combination: 2C-H stretching and C-H deformation
6010	1664	cis R$_1$CH=CHR$_2$	cis C-H vibration
5790	1727	–CH$_2$–	C-H 1st overtone
5678	1761		C-H 1st overtone
5260	1901	C=O	C=O stretching 2nd overtone
5178	1931	C=O str	C=O stretching 2nd overtone
4708	2124	-COOR	C-H stretching and C=O stretching
4662	2145	–CH=CH–	Combination: =C-H stretching and C=O stretching
4596	2176	–CH=CH–	C-H stretching asym and C=C stretching
4329	2310		C-H combinations and deformation
4255	2350		C-H combinations and deformation

The low-intensity bands originating from second overtones of the C-H stretching vibrations are present at around 8261 cm^{-1} and C-H combinations with other modes at 7180 and 7075 cm^{-1} (Sinelli et al. 2010a).

Raman Spectra

Raman spectra of vegetable oils contain well-defined bands with positions similar to those in MIR spectra (Fig. 7.3, Table 7.4).

The relative intensities of the particular bands are very different from those in MIR spectra. For example, the bands originating from vibrations of polar C=O group in ester linkage have a very weak intensity in Raman spectra. In contrast, the band near 1660 cm^{-1} that corresponds to the vibration of nonpolar C=C groups is intense.

The 1700–1600 cm^{-1} range is important to measure the degree and type of unsaturation of oils. The exact position of the intense bands corresponding to the C=C stretching vibration depends on configuration of the double bonds; the bands occur at 1656 and 1670 cm^{-1} for *cis* and *trans* bonds respectively (Van de Voort et al. 2001). Generally, IR absorption and Raman scattering spectra provide complementary information (Van de Voort et al. 2001; El-Abassy et al. 2009; Korifi et al. 2011; Zhang et al. 2011).

Vibrational and Electronic Spectroscopy and Chemometrics in Analysis of Edible Oils 209

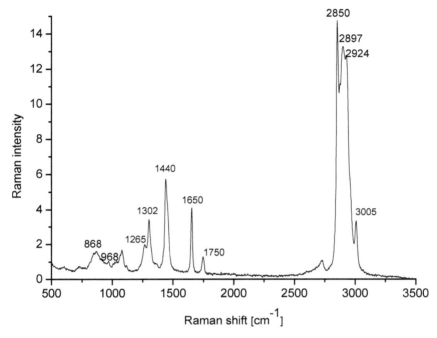

Figure 7.3 Raman spectrum of olive oil.

Table 7.4 Functional groups and modes of vibrations in oils Raman spectra (El-Abassy et al. 2009).

Theoretical cm^{-1}	Functional group	Mode of vibration
3005	cis RHC=CHR	=C-H stretching sym
2924	–CH$_2$	C-H stretching sym
2897	–CH$_3$	C-H stretching sym
2850	–CH$_2$	C-H stretching sym
1750	RC=OOR	C=O stretching
1650	cis RHC=CHR	C=C stretching
1525*	RHC=CHR	C=C stretching
1440	–CH$_2$	C-H bending (scissoring)
1300	–CH$_2$	C-H bending (twisting)
1265	cis RHC=CHR	=C-H bending (scissoring)
1150*	–(CH$_2$)$_n$–	C-C stretching
1008*	HC–CH$_3$	CH$_3$ bending
968	trans RHC=CHR	C=C bending
868	–(CH$_2$)$_n$–	C-C stretching

*bands attributed to carotenoids

7.2.3 Electronic Spectroscopy of Oils

The UV spectral range spans the 200 to 350 nm wavelength interval, and the Vis range—350 to 800 nm. The photon energy corresponds to an electronic transition involving electrons engaged in chemical bonds in a molecule. Based on the type of interaction of radiation with molecules, the electronic spectroscopy is classified as absorption or emission. Absorption of a photon raises a molecule to electronic excited states. Deactivation of an excited molecule may occur by radiationless transition or by photoluminescence—emission of a photon with a different energy. Fluorescence is photoluminescence. It involves emission of a photon due to a transition between the excited and ground states of the same multiplicity, occurring without changes in the electronic spin, usually $S_1 \rightarrow S_0$. Fluorescence is usually observed in organic aromatic compounds with conjugated double bonds and rigid molecular skeletons (Lakowicz 2006).

Ultraviolet and Visible Absorption Spectra of Oils

Edible oils exhibit intense absorption in the UV range attributed to compounds with groups with high electronic density, such as carbonyl groups, nitro groups, double and triple bonds, conjugated double bonds, etc.

Unsaturated fatty acids that contain isolated *cis* double bonds absorb UV light at the wavelength close to 190 nm. This wavelength is nonspecific, since a great number of compounds have an absorption in this region. Conjugated fatty acids absorb at various wavelengths in the UV range, depending on the length of conjugation and configuration of the double-bond system (Belitz et al. 2009). The presence of conjugated dienes in fatty acids results in a characteristic absorption at around 234 nm. Conjugated-triene fatty acids that may also be formed in the refining treatment, exhibit three absorption maxima at around 259, 268 and 279 nm. The specific absorption coefficients measured at 232 nm and 270 nm (K_{232} and K_{270}) are used as official indicators for olive oil quality control, including detection of oil oxidation products and adulteration with refined oils.

Different chromophores are present in fatty acids with one double bond conjugated with a carboxyl group; they have absorption maxima near 208–210 nm and have an extinction coefficient lower than that of the conjugated dienes.

Absorption spectra of crude vegetable oils in the visible region exhibit bands corresponding to liposoluble pigments (Fig. 7.4).

The main carotenoid pigments in olive oils are either β-carotene or lutein, and the main pigment of chlorophyllic groups is pheophytin *a* (Moyano et al. 2008).

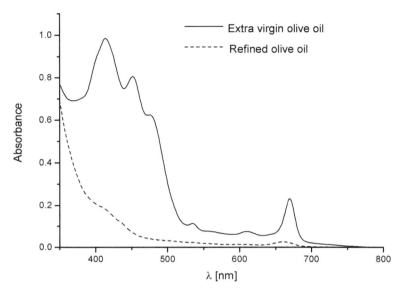

Figure 7.4 Visible absorption spectra of olive oils.

The absorption in the 450–520 nm range originates from carotenoid pigments. The absorption bands of carotenoids overlaps with the chlorophyll absorption at 380–450 nm, while the characteristic band at 650–700 nm originates exclusively from the absorption of chlorophylls. Refined oils reveal only traces of pigment absorption as these are removed or degraded during refining.

Fluorescence Spectra of Oils

Fluorescence spectra of oils provide enhanced selectivity as compared to the absorption spectra. The most comprehensive characterization of the multicomponent fluorescence of oils is obtained by measurement of an excitation-emission matrix, or the total luminescence spectrum (Ndou and Warner 1991; Guilbault 1991) or synchronous fluorescence techniques, proposed by Lloyd (1971).

Total fluorescence spectra of edible oils (Fig. 7.5) show bands characteristic for tocopherols, with the maximum at excitation/emission wavelengths ($\lambda_{exc}/\lambda_{em}$) at about 300/331 nm, and pigments of the chlorophyll group with the maximum at $\lambda_{exc}/\lambda_{em}$ at about 400/680 nm (Zandomeneghi et al. 2005; Diaz et al. 2003). The emission at shorter wavelengths also has contributions of phenolic compounds, while the bands ascribed to the oxidation products are observed for some oils in the intermediate region.

212 Methods in Food Analysis

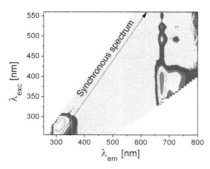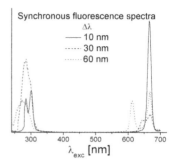

Figure 7.5 Total fluorescence and synchronous fluorescence spectra of an olive oil.

In the synchronous fluorescence spectra (Fig. 7.5, recorded at the emission-excitation offset of 10 nm), the bands corresponding to tocopherols and chlorophylls appear with their maxima at 301 and 666 nm, respectively. An additional band at 284 nm may be ascribed to the emission of phenolic compounds (Sikorska et al. 2012). The bands observed for some oils in the 320–500 nm range are ascribed to the oxidation products.

7.3 Chemometric Analysis of Spectra of Edible Oils

7.3.1 General Ideas

Among many definitions of chemometrics, the one that best explains its importance in the analysis of spectroscopic data seems to be that proposed by the International Chemometrics Society. Chemometrics is defined as "the science of relating measurements made on a chemical system or process to the state of the system via application of mathematical or statistical methods" (Hibbert et al. 2009).

The spectra are characteristic for each particular sample. When pure chemical compounds are studied, spectra provide information about their molecular structure and properties, and interactions with surroundings molecules. The spectroscopic techniques are powerful research tools applicable to a diversity of molecular systems. When the spectra are recorded for food samples, they contain information about different components of the sample matrix and their interactions. Although this may make the direct interpretation difficult (without precluding it), the spectra provide overall characteristics of a complex food product, constituting its fingerprint. Multivariate methods are normally used in interpreting the spectra for analytical purposes, and are especially suitable for systems described by many variables (Bro 2003; Kjeldahl and Bro 2010).

The analysis of spectra using multivariate methods provides qualitative and quantitative information about the samples under study. Multivariate methods are used for exploratory analysis of spectra, to detect any patterns in the data set. The main objectives of the multivariate methods in the spectra analysis are to find patterns in the data, classify the samples, and model the relationships between the spectra and the evaluated properties. An important field is application of spectroscopy coupled with chemometrics to process analytical technology (PAT). These methods can be used not only to control raw materials, intermediates and final products, but also to monitor and to model processes (Moros et al. 2010; Armenta et al. 2010b; Marquez et al. 2005).

7.3.2 Exploratory Analysis (PCA, CA)

Exploratory data analysis is aimed at discovering patterns in data sets: similarities and differences, clustering of samples, any systematic trends and outlier detection. This analysis does not require any prior knowledge of the explored data and is performed employing unsupervised pattern recognition methods. One of the most popular and widely used methods is the Principal Component Analysis (PCA). The basic idea behind PCA is a reduction of the data by removing useless information (noise and redundancy) while retaining meaningful analytical information.

Principal component analysis. Principal component analysis is a variable-reduction method. It transforms the original variables into a smaller number of new uncorrelated variables—principal components (PCs). A mathematical model is constructed according to the equation:

$$X = TP^T + E \qquad \text{eq. (7.1)}$$

Where X is the original data set consisting of N rows (samples) and M columns (variables), A is the number of principal components, T (N×A) is the scores matrix, P (M×A) is the loading matrix, and E (N×M) the residuals matrix of the model.

The principal components are determined on the basis of the maximum variance criterion. They are linear combinations of the original variables and are orthogonal to each other. The optimal number of PCs is estimated based on several criteria: percentage of the explained variance, eigenvalue-one criterion, screen test, and cross-validation. The number of PCs that is required to explain most of the variance of the original data is usually much smaller than the number of original variables.

Interpretation of the results of PCA is usually carried out by visualization of the component scores and loadings (Otto 2007). The PC scores are usually plotted as two-dimensional plots, where the axes are defined by different (usually the first two) principal components. Such plots are used to assess the relationships between the samples. The loadings are used to assess relationships between different variables in the model.

Principal component analysis (PCA) is probably the most frequently applied multivariate method, and is widely used for data screening and compressing multivariate data.

Cluster analysis. Cluster analysis is another unsupervised pattern recognition technique. Using this method, samples are grouped into clusters based upon the similarities between them, with no additional knowledge about class membership required. Distance, correlation or combination of the two may be used as the measure of similarity. Depending on the criteria used to define the distance between the two groups (linkage rule), different clustering algorithms are used to group the samples: single (nearest neighbor), complete (furthest neighbor) or average linkages, centroid method, Ward's method, etc. As a result, samples within a particular cluster are more similar to each other than to those outside that cluster (Otto 2007).

The exploratory methods are normally used as the first step in data analysis and provide important information. However, the most desired application of multivariate methods is development of a model that describes relations between particular spectra and sample properties. Once such a model is established and validated, it may be used to predict the properties of interest based on the spectra. It is especially attractive for parameters whose determination is difficult and requires extended chemical procedures. In such cases chemical methods may be replaced by much easier, direct spectroscopic measurements.

7.3.3 Multivariate Quantitative and Qualitative Models

Constructing of multivariate classification or calibration model is a complex process that involves several steps, summarized in Table 7.5.

Table 7.5 Steps in multivariate model development.

Step	Calibration	Classification
Selection of the calibration sample set	+	+
Determination of target parameters using reference methods	+	-
Acquisition of spectra	+	+
Mathematical preprocessing of data	+	+
Construction of the model	+	+
Validation of the model	+	+
Prediction of unknown samples	+	+

Selection of the Calibration Sample Set

This is the first and one of the most important steps in developing a calibration model. The set of samples selected to develop the calibration model should be large enough and representative of samples that will be analyzed. ASTM International standard suggest a 6:1 ratio of samples to variables (Small 2006). Samples should represent a wide range of values of the parameters of interest, and should cover all possible sources of variation. The correlations among the concentrations of the components must be minimized. Extreme samples should be included into this set. The calibration samples have to be handled in the same way as new unknown samples will be handled in subsequent analyses.

Several algorithms are used for efficient selection of calibration samples: D-optimal, Duplex, OptiSim, Kennard and Stone (Moros et al. 2010).

Determination of Target Parameters

When the classification model is developed, the knowledge about membership of the samples in a certain category is used, thus usually there is no need to perform the analysis using the reference method.

Development of a calibration model for the quantitative prediction of the parameters always requires the determination of the parameters of interest with adequate accuracy and precision. It is important to assess the figures of merit of the methods used to obtain the target parameters, since the analytical performance of the reference procedure is crucial to the performance of the calibration model. The error associated with the reference method (standard laboratory error) should be known. Note that no chemometric techniques can compensate for poorly designed experiments or inadequate experimental data.

Acquisition of Spectra

The performance of the calibration depends also on the quality of collected spectra. The optimal measurement mode has to be evaluated depending on the physical state of the sample. Transparent liquid samples are usually measured in the transmission mode. The reflection measurements are used for (optically) infinitely thick samples. Diffuse reflection is used in NIR and attenuated total reflection in MIR region. Fluorescence of diluted samples is measured with right-angle geometry, for samples with high absorbencies front-face alignment is used.

The order of spectra collection should be randomized to minimize the correlations between concentrations and time (Small 2006).

The diagnostics of the instrument has to be checked prior to the analysis, including: wavelength accuracy, signal to noise ratio, and repeatability. The spectra of the calibration samples and those used for future prediction should contain the same sources of instrumental variability; therefore, all of the spectra should be obtained using the same instrument and measurement conditions.

Mathematical Pre-processing of Spectra

The collected spectra are usually mathematically preprocessed to enhance the analytical information and remove those spectral variations that are unrelated to the sample components or properties of interest. The aim of preprocessing is to improve the linear relationship between the spectral signals and the analytic concentrations. This leads to the simplest possible calibration model that enables prediction of new samples with good precision and accuracy.

The preprocessing methods are chosen using different approaches: trial and error, visual inspection of data, and assessment of pre-processed data by quality parameters, to quantify the presence of artifacts. The choice of the optimal pre-processing methods is of great importance as it may strongly influence the final result (Engel et al. 2013).

The preprocessing of spectra may be divided into two types: spectral scaling applied to each individual variable in all samples (mean centering, and auto-scaling) and spectral preprocessing applied to individual spectra (derivative spectra, standard normal variate, and multiplicative scattering correction). Some of the frequently used preprocessing techniques are outlined below.

Mean centering. Mean centering relies on the subtraction of the mean response of the variable over all of the samples in the data set from each individual variable response. As a result, the absolute intensity information is removed from the data, the remainder being interpretable in terms of variation around the mean (Nicolaï et al. 2007).

Auto-scaling. This involves initial mean centering of the data followed by division of the resulting intensities of each variable by its standard deviation. Each of the variables in autoscaled data has a zero mean value and a standard deviation of one. As a result, each of the variables has equal influence on the model. This procedure is normally used when different variables are measured using different types of instruments, have different units or different ranges.

Differentiation. Application of the first derivative in spectroscopy effectively reduces the baseline offset between samples. The second derivative removes both the baseline offset and the differences in baseline slope between spectra. The most frequently used technique for derivation of spectra is the Savitzky-Golay routine. The derivatives are evaluated by fitting a polynomial expression to the data within a window moving along the wavenumber or wavelength axis. The procedure requires the number of points in the window and the order of polynomial as input parameters. The Savitzky-Golay algorithm may be additionally used for spectral smoothing and noise reduction, causing some reduction in spectral resolution.

Multiplicative scatter correction. It is used for the compensation of non-linearity induced in spectral data by physical scattering processes. The degree of scattering is dependent on the radiation range, size of the particles and the refractive index. The MSC allows compensating for additive effects, such as baseline shift, and multiplicative effects, such as tilt in the spectral data. It is performed by estimating the correction coefficients and applying the correction. A limitation of the MSC is that it assumes that offset and multiplicative effects are much larger than those caused by chemical changes in a sample. Extensions of MSC were proposed that also allow compensating for chemical interferences by using known spectra of analytes and interferents.

Standard normal variate (SNV). Similarly to MSC, it reduces additive and multiplicative effects, both techniques giving similar result. They differ in the way of calculating the correction factors.

Variable Selection

An important step in multivariate analysis is the selection of the variables (spectral intensities in wavenumber or wavelength ranges) that contribute useful analytical information and elimination of the variables containing mostly noise. The identification of the appropriate spectral ranges reduces the complexity of the model and improves its predictive ability. Classically the selection of variables has been made manually based on knowledge about spectroscopic properties of the studied samples. However, it has been shown that objective mathematical strategies are more efficient. The available selection methods include: successive projections algorithm (SPA), uninformative variable elimination (UVE), simulated annealing (SA), artificial neural networks (ANN), genetic algorithms (GAs), interval partial least squares (iPLS), windows PLS, iterative PLS and wavelength selection with B-spline, Kalman filtering, Fisher's weights and Bayesian approach (Xiaobo et al. 2010).

Construction of Models

Quantitative Models

The aim of the calibration process is to establish a mathematical relationship between spectra and properties of the studied samples. When the intensity of the analytical signal depends exclusively on the concentration of the analyte (or other property), usually a straight line is used as a calibration curve. However, in the analysis of complex food samples by spectroscopic techniques the analytical signal usually comprises contributions of interferences or matrix effects. Thus, multivariate calibration methods have to be used for quantitative analysis of spectra. Principal component regression and partial least square regression are the multivariate methods most widely used in analytical chemistry. Both of these approaches rely on the reduction of the number of variables, helping to overcome the problems of poor selectivity and collinearity.

Principal component regression is a two-step procedure based on PCA and multiple linear regression (MLR) methods. In the first step, the data matrix X is decomposed using PCA methods. The obtained scores matrix T is next used as the predictors X matrix in MLR. The advantage of this approach with respect to MLR is that PCs used as the X-variables are uncorrelated, and that the noise is filtered. A small number of PCs is usually sufficient. A disadvantage is that the principal components are calculated without taking into account the Y matrix variance. As a result, the variance explained by the principal components in some cases may be irrelevant for predicting the Y value (Wold et al. 2001).

Partial least square regression is the method most often used for analysis of spectral data. Partial least squares regression (PLSR) is a multivariate projection method for modeling a relationship between dependent variables (Y) and independent variables (X). PLSR uses the information contained in both the spectral data matrix (X) and the concentration vector (Y) during the calibration. The aim of the analysis is to find such latent variables in the input matrix (X) that describe most of the relevant variations in this matrix and at the same time predict the latent variables in Y. PLSR maximizes the covariance between the spectral matrix X and the response vector Y. Less weight is given to the variations that are irrelevant or noisy. The latent variables are ordered according to their relevance for predicting the Y-variable.

PLS can be used for analysis of data strongly correlated, noisy and redundant, and also allows modeling and predicting several Y variables at a time, using the PLS2 algorithm. The optimal PLS model typically uses a smaller number of relevant latent variables than a PCR model with a similar performance.

Qualitative Models

The aim of classification, achieved by qualitative models, is to assign new samples to previously defined categories or classes by using their measured (spectral) data. Classification models are developed using a training set of samples of known membership to a particular category and employing supervised pattern recognition techniques.

The methods used for classification may be grouped into discrimination and class modeling categories. Using the discriminating techniques, the hyperspace of sample parameters is divided into as many regions as the number of available categories, and each of the samples is always classified into only one of the defined classes. The discrimination methods are: linear discriminant analysis (LDA), *k*-nearest neighbors (*k*NN), partial least squares discriminant analysis (PLS-DA), classification and regression trees (CART), and artificial neural networks (ANN).

The class modeling techniques are based on modeling the analogies between samples that belong to a defined category. Each class is modeled separately, and boundaries are defined that separate a particular class from the rest of the hyperspace (Sun 2009). Using these methods, a particular sample may be classified into one category, into more than one category, or into none of the defined classes or categories.

The class modeling methods are: soft independent modeling of class analogy (SIMCA) and unequal dispersed classes (UNEQ) (Oliveri and Downey 2012).

Linear discriminant analysis is the first multivariate classification technique developed and one of the frequently used and well-studied. LDA considers both within-group variance and between-group variance. It relies on classification of objects into classes by maximizing the ratio of between-class variance to the within-class variance. It determines linear discriminant functions, called canonical variates, which are linear combinations of the original variables and are orthogonal to each other (Bakeev 2005). The number of canonical variates for n classes is equal to $n-1$, if the total number of variables is larger than n.

LDA, similarly to PCA, is a feature-reduction method. However, the canonical variates show the direction in space that maximizes the ratio of between-class variance to within-class variance, while the principal components maximize the between-object variance, disregarding any information about class membership of the samples. As a consequence, LDA is better adapted to discriminating between classes, and in some cases fewer LDs than PCs are needed to provide appropriate discrimination on the same data (Bakeev 2005).

LDA assumes the normal distribution of classes. Moreover, it is assumed that the covariance matrices of the two (or more) classes are identical; in other words, the variability within each group has the same structure. LDA requires that the number of samples is higher than the number of variables. Highly correlated variables induce the risk of overfitting. Discriminant analysis always assigns objects to one of the classes. LDA is a variant of discriminant analysis, in which the discrimination boundaries are linear (Berrueta et al. 2007). Other functions can be used for discrimination, such as quadratic discriminant function (QDA) and Bayesian classification function, which are sub-cases of regularized discriminant function (RDA).

K-nearest neighbor (kNN) method is a non-parametric technique frequently applied in classification. This method determines the distances (Euclidean or other) between an unknown sample and each of the objects in the training set. Next, *k* objects nearest to the unknown sample are selected, and using the majority rule, the sample is classified into the group to which the majority of the *k* objects belong.

The optimal value of *k* is chosen by calculating the prediction ability with different *k* values.

Typically small *k* values (1, 3 or 5) are used (Berrueta et al. 2007; Otto 2007). Prior to analysis, data are mathematically preprocessed to avoid the effect of different scales of variables.

Mathematical simplicity is an advantage of this method. It may be used even if only a few calibration samples are available. Furthermore, it does not assume any specific distribution of the variables or the linear separation of the classes.

The limitations of *k*NN are that is provides no assessment of confidence in the classification results. Moreover, there is poor information about the structure of the classes and of the relative importance of variables in the classification. It does not allow graphical representation of the results (Berrueta et al. 2007).

Soft Independent Modeling of Class Analogy (SIMCA) is the most used class-modeling technique. The method is based on independent, separate PCA modeling of each class in the calibration set. Each class model can be described by a different number of significant principal components that are determined by cross-validation. Thus, an optimal number of PCs is estimated that accounts for most of the within-class variance. On the other hand, high signal-to-noise level is assured by excluding the so-called noise-laden principal components from the model. The unknown sample is assigned to a particular class based on its analogy with the calibration samples that are members of this class. Classification of unknown samples is achieved by comparison of their distances from each of the class models.

SIMCA provides information about the class distance and the modeling and discriminatory powers and its results may be presented graphically. The discriminatory power informs how well a particular variable discriminates between the two classes. The modeling power indicates variables being able to model a particular class well.

Partial least-squares discriminant analysis. This approach uses quantitative regression methods of PLS to perform qualitative analysis (Bakeev 2005). The "dummy" Y matrix contains either zero or one value, depending on the class membership of the calibration samples. An Y variable has the value 1 for a sample that belongs to the class and zero for the remaining samples. The number of Y variables is determined by the number of classes. In case of multiple Y variables, the PLS2 method is used. The X matrix contains original or preprocessed data. The aim of the analysis is to predict one (or several) binary responses Y from a set of variables in X. The X and Y matrices are decomposed into a product of other two matrices, matrix of scores and matrix of loadings. The resulting scores not only retain the maximum variances of the original data but also are correlated with the class-variable.

An optimal number of latent variables is estimated by using cross-validation or external test sets.

PLS takes into account the errors in both X and Y matrices, assuming that they are equally distributed. Moreover, PLS is suitable for data sets with number of samples fewer than number of variables, and is tolerant to high degrees of inter-correlation between independent variables.

Validation of the Model

An important issue in developing calibration or classification models is validation. It involves the evaluation of model predictive ability for unknown samples that were absent in the calibration set (Sun 2009).

Different procedures may be used for model validation. In external validation independent sample sets are used for model developing (training or calibration set) and validation (test set). Each of the two sets should contain representative samples.

Cross-validation relies on developing a model using only part of data. The remaining data are then used for model testing. Several cross-validation procedures are available for model validation, including: selected subset, leave-one-out, random cross-validation, block-wise cross-validation, and alternating-sample cross-validation (Bakeev 2005).

The final validation for any calibration model is to assess its performance when applied to data collected outside the time frame of the calibration.

The validation of classification models involves estimation of the recognition and prediction ability. The recognition ability is percentage of the samples in the training set correctly classified during modeling. The prediction ability is the percentage of samples in the test set correctly classified by the developed model. The quality of a classification model is evaluated by its sensitivity and selectivity. The sensitivity of a model is the proportion of samples belonging to a particular class that are correctly classified into this class. Model selectivity is the proportion of samples not belonging to the class that are classified as foreign (Berrueta et al. 2007).

The quality of the regression model is evaluated by means of several parameters. The prediction performance is estimated by the root mean square error for cross validation (RMSECV) or root mean square error for prediction, when respectively internal or external validation is used:

$$\text{RMSEP} = \sqrt{\frac{\sum_{i=1}^{N}(\hat{y}_i - y_i)^2}{N}}, \qquad \text{eq. (7.2)}$$

Where: N is the number of validated samples, $\hat{y}_i$ and y_i are the model-predicted and measured value for the external validation sample i, respectively (Bakeev 2005). RMSEP and RMSECV express the average expected uncertainty for new samples and are given in the units of property of interest. These values are used to assess complexity of the calibration model. The number of latent variables used in a regression is estimated so as RMSEP or RMSECV are minimized.

Other figures of merit used to estimate quality of calibration models are: the correlation between predicted and actual values, expressed by the multiple coefficients of determination (R^2); the systematic averaged deviation between the data sets of the actual and predicted values (bias).

A properly validated model may be used for prediction of new samples. The information necessary for interpretation of a multivariate calibration model is summarized in Table 7.6.

7.4 Application of Spectroscopy and Chemometrics in the Analysis of Edible Oils

Tables 7.7 and 7.8 list the applications of spectroscopy coupled with multivariate methods in qualitative and quantitative analysis of edible oils. The examples were chosen to illustrate the wide range of application of these techniques. The objective is not a comprehensive review of the literature on this subject.

Table 7.6 Information required for interpreting a multivariate calibration model that must be included (Cozzolino et al. 2011).

Sample structure	Number of samples
	Range and Standard deviation of reference parameters
Reference method	Method-procedure
	Limit of detection
	Standard error of the laboratory
Model statistics	Coefficient of determination in calibration
	Slope
	Bias
	RMSECV, RMSEP
	Number of terms or principal components
	Number of samples
	Number of samples removed as outliers

The range of quality aspects of oils studied by spectroscopy-chemometrics is wide, and includes application of both calibration and classification models. Vibrational and electronic spectroscopy coupled with multivariate regression analysis were employed for quantitative determination of chemical and physical oils properties. Vibrational spectroscopy was used to determine major and minor oil components. Fresh oils were analyzed for fatty acid and triacyloglycerol profiles. Analysis of minor components in the unsaponifiable matter includes determination of such analytes as chlorophylls, carotenoids, tocopherols, and phenolic compounds. The concentration of polymerized triacyloglycerols in oxidized oils may be predicted using spectra. Chemical parameters used for evaluation of oil quality: peroxide, anisidine, iodine values, free fatty acids were determined. Physical properties like refractive index, viscosity and relative density may be predicted using spectra. Organoleptic properties, flavor, aroma and color may also be predicted from the respective spectra.

A number of spectroscopic techniques coupled with multivariate methods have been exploited to classify or discriminate oils according to different criteria. Spectroscopy was used to identify the botanical origin of oils. Classification of the samples of olive oils of different quality grades may be performed based on spectral properties.

Spectroscopic techniques have been used to evaluate and quantify the adulteration of extra virgin olive oils with lower quality olive and other vegetable oils. They enabled discrimination between oil samples of

Table 7.7 Examples of applications of vibrational and electronic spectroscopic techniques in quantitative determination of oil components.

Spectroscopic Technique	Components or parameters to be determined	Samples	Multivariate methods	References
NIR transmission	phenolic compounds, degree of unsaturation (fatty acid profile), hydroxytyrosol derivatives and C6 alcohols, sensory attributes	virgin olive oils, n=97	PLS	(Inarejos-García et al. 2013)
NIR transmission	fatty acids, triacylglycerols	virgin olive oil samples, n=125	PLS	(Galtier et al. 2007)
NIR transmission, flow-cell	acidity, refractive index and viscosity	vegetable oils: corn, soya, canola and sunflower, n=70	iPLS, (SPA) and (MLR)	(Pereira et al. 2008)
NIR transmission	polymerised triacylglycerols in deep-frying vegetable oils	olive, sunflower, corn and seed oil; deep fried oils in the absence and presence of foodstuff	PLS (iPLS), (UVE-PLS)	(Kuligowski et al. 2012)
NIR transmission	total polar materials and free fatty acids in frying oils	partially hydrogenated soy-based frying, different levels of degradation, n=60	FSMLR; PLS	(Ng et al. 2006)
NIR fiber-optic probe with a liquid attachment	fatty acids with emphasis on trans fatty acids	canola, corn, flax, olive, soybean, sunflower, walnut oils, margarines, shortenings, lard, partially hydrogenated oil fractions (from canola and soybean oil)	PLS	(Azizian and Kramer 2005)
NIR transmission	free fatty acids, peroxide value, polyphenol content, induction time, chlorophyll, major fatty acids	olive oil, n=216	PLS	(Mailer 2004)
MIR ATR	total amount of methyl and ethyl esters of fatty acids, the ratio between ethyl and methyl esters	extra virgin olive oils, n=81	PLS	(Valli et al. 2013)
MIR ATR	total amount of chromanols (tocopherols, tocotrienols and plastochromanol-8)	canola, flax, soybean and sunflower seeds, n=17	PLS	(Ahmed et al. 2005)

MIR transmission	refraction index, relative density	canola, sunflower, corn, soybean oil, n=103	PLS, SVM	(Luna et al. 2013a)
MIR ATR	water content, total phenol and antioxidant activity	virgin olive oil and olive oil, n= 47	PLS	(Cerretani et al. 2010)
MIR ATR	evaluation the impact of fly (*Bactroceraoleae*) attack on olive oil quality	virgin olive oils, n=32	PLS	(Gómez-Caravaca et al. 2013)
MIR ATR	fatty acid profile (oleic, linoleic, saturated, mono-unsaturated, poly-unsaturated fatty acids),peroxide value	virgin olive oil, n= 86	PLS	(Maggio et al. 2009)
MIR ATR	oxidized fatty acids	virgin olive oil, n=72	MLR	(Lerma-García et al. 2011)
MIR single bounce attenuated total reflectance (SB-ATR)	erucic acid	rapeseed, canola, sunflower (mixtures), n=13	PLS	(Sherazi et al. 2013)
UV-vis, NIR transmission MIR ATR	oleic and linoleic acids	extra virgin olive oils, n=57	PCA, PLS	(Casale et al. 2012)
Raman (CRS) Vis	fatty acid and triacylglycerol	virgin olive oil samples, n = 396	PLS	(Korifi et al. 2011)
Raman Vis	carotenoids	heat-induced degradation of extra virgin olive oil	PLS	(El-Abassy et al. 2010)
Raman	adulteration of extra-virgin olive oil with sunflower oil	binary mixtures of extra-virgin olive oil	PLS	(El-Abassy et al. 2009)

Table 7.7 contd....

Table 7.7 contd.

Spectroscopic Technique	Components or parameters to be determined	Samples	Multivariate methods	References
Fluorescence, synchronous	adulteration of extra-virgin olive oil with olive oil	extra virgin olive oil, olive oil		(Dankowska and Malecka 2009)
Fluorescence excitation-emission	K232, K270, peroxide value	extra virgin, virgin, pomace olive oils, n=33	PARAFAC, N-PLS	(Guimet et al. 2005b)
Fluorescence front face and right angle synchronous	total tocopherol content	olive, grape seed, rapeseed, soybean, sunflower, peanut, and corn oils, n=25	PCA, PLS	(Sikorska et al. 2005)

iPLS—interval partial least squares, SPA—the successive projections algorithm, MLR—multiple linear regression, FSMLR—forward stepwise multiple linear regression, WT—wavelet transform, UVE—elimination of uninformative variables, SVM—support vector machines, CRS—confocal Raman spectroscopy, PARAFAC—parallel factor analysis, N-PLS—multiway partial least-squares regression.

Table 7.8 Examples of applications of vibrational and electronic spectroscopic techniques in classification and discrimination analysis of oils.

Spectroscopic Technique	Discrimination/classification criteria	Samples	Multivariate method	References
NIR transmission	non-transgenic and transgenic soybean oils	soybean oils, n=80	PCA, SIMCA, SVM-DA, PLS-DA	(Luna et al. 2013b)
NIR transmission MIR ATR	sensory olfactory attributes	virgin olive oils, n=112	LDA, SIMCA	(Sinelli et al. 2010b)
MIR ATR NIR transmission Raman	discrimination among edible oils and fats	butter, lard, cod liver extra virgin olive, corn, peanut, canola, soybean, safflower, coconut, n=110	PCA, PLS, LDA, CVA	(Yang et al. 2005)
NIR transflectance MIR ATR	authentication of the PDO extra virgin olive oil	extra virgin olive oil PDO area of Sabina, n=20, from other origins (Italy or Mediterranean countries), n= 37	PLS-DA, SIMCA	(Bevilacqua et al. 2012)
UV–Vis transmission NIR transmission MIR ATR	authentication of the PDO extra virgin olive oil	extra virgin olive oils n=57	UNEQ, SIMCA	(Casale et al. 2012)
MIR HATR	detection adulteration of extra-virgin olive oil with vegetable oils	extra virgin olive oil rapeseed, cottonseed and corn–sunflower binary mixture	PCA, PLS-DA	(Gurdeniz and Ozen 2009)
MIR transmission	discrimination between different oils	corn, canola, sunflower, soya, olive, butter, n=255	PLS-DA, iPLS-DA, ECVA, iECVA	(Javidnia et al. 2013)
MIR ATR	differentiation between blends of edible oils with olive oil content higher than and below 50% (w/w)	olive, canola, corn, flaxseed, grape seed, peanut, rapeseed, safflower, sesame, soybean, sunflower, high oleic sunflower oils	PLS-DA	(De la Mata et al. 2012)
MIR ATR	botanical origin, composition of binary mixtures of extra virgin olive oil with other low cost edible oil	extra virgin olive oil, sunflower, corn, soybean and hazelnut)	LDA	(Lerma-García et al. 2010)

Table 7.8 contd....

Table 7.8 contd.

Spectroscopic Technique	Discrimination/classification criteria	Samples	Multivariate method	References
MIR ATR	distinguishing geographic origin of extra virgin olive oils	extra virgin olive oils from Italy, Greece, Portugal, Spain, n=60	PLS and LDA GA and LDA	(Tapp et al. 2003)
MIR ATR Raman	detection of the adulteration of refined olive oil with refined hazelnut oil	refined, lampante, virgin olive oils, hazelnut oils (refined and crude), n= 233	Stepwise LDA	(Baeten et al. 2005)
Raman	distinguishing between closely related cultivars of extra virgin olive oils and hazelnut oils	extra virgin olive oil, hazelnut oils	PLS genetic programming (GP)	(López-Díez et al. 2003)
UV-Vis	to quantify adulterations of extra virgin olive oil with refined olive oil and refined olive-pomace oil	extra virgin olive oil, n= 396	chaotic parameters (Lyapunov exponent, autocorrelation coefficients, and two fractal dimensions, CPs)	(Torrecilla et al. 2010)
Fluorescence excitation-emission	detection adulteration in PDO olive oil	virgin olive oils from the PDO "Siurana" n=34	unfold-PCA, PARAFAC, LDA, discriminant N-PLS	(Guimet et al. 2005a)
Fluorescence excitation-emission	discrimination between quality grades of olive oils	virgin olive, pure olive, olive-pomace oils, n=56	CA	(Guimet et al. 2004)
Fluorescence Total synchronous	categorizing edible oils according to thermal stress	extra virgin olive, olive pomace, sesame, corn, sunflower, and soybean oils and a commercial blend of oils, heated at 100, 150, and 190 °C	PCA	(Poulli et al. 2009)

SVM-DA—Support Vectors Machine-Discriminant Analysis, PLS-DA—Partial Least Squares-Discriminant Analysis, GA—a genetic algorithm, CVA—canonical variate analysis, UNEQ—unequal class models, SIMCA—soft independent modeling of class analogy, HATR—horizontal attenuated total reflectance, ECVA—extended canonical variate analysis, GP—genetic programming.

very similar and geographically close denominations of origin. Recently spectroscopy was used for discrimination of non-transgenic and transgenic soybean oils. It is possible to classify oils according to sensory olfactory attributes using spectra.

Most published studies utilized vibrational spectroscopy in the infrared region. These spectra contain structural information about sample components. The largest number of applications involves NIR spectroscopy. The advantage of this technique is higher penetration depth of radiation—near-infrared light penetrates much deeper than MIR into an intact food sample (>10 mm), and usage of fiber-optic technology in remote sensing. The drawback of NIR is the less interpretative character of spectra as compared to the MIR, due to bands overlap and lower intensity. The sensitivity of this technique to minor constituents is low; however, it depends on the chemical characteristics of the analyte and the complexity of the sample matrix studied. NIR spectroscopy is most extensively used in practical real-world applications. Commercial analyzers using NIR techniques are available for oil analysis. They are utilized for determination of different parameters: moisture, fatty acid content, iodine index, phosphorous value and others (Armenta et al. 2010a).

Recently an increase of Raman spectroscopy applications has been observed. This technique provides spectral information complementary to the MIR spectroscopy, and measurements may be conducted using fiber optics. Raman and NIR spectroscopies allow the analysis of oils through packaging material, such as plastic or glass.

Electronic spectroscopy coupled with chemometrics has been used in published oils studies to a lesser extent. Often the absorption spectroscopy in the VIS region is used together with NIR spectroscopy, as some instruments enable acquisition of spectra in both regions. Applications of methods based on fluorescence recently became more frequent as well. This technique is characterized by enhanced sensitivity and selectivity as compared to absorption techniques. This enables the analysis of minor components in a complex matrix.

The choice of an appropriate spectroscopic technique is dependent on the analytical problem studied and determined by the required spectral information, sample and analyte characteristics, and available sampling options.

7.5 Conclusion

The numerous spectral bands present in oil spectra in different radiation regions are associated with the structure of oil molecular components and their characteristics, and are affected also by physical properties, therefore the spectral pattern is unique for the particular sample. Chemometric analysis of the spectral data enables extraction of analytically useful information and establishment of calibration and classification and/or discrimination models. Such models allow prediction of a variety of oil properties based on the spectral measurements.

One of the most important features of spectroscopy-chemometrics methods in quality control applications is the possibility of performing non-destructive measurements directly on the untreated samples, thus avoiding time- and labor-consuming chemical treatment steps. Elimination of chemical sample pretreatments or physical separations reduces or eliminates the usage of reagents or solvents and reduces the amount of environmentally harmful waste. The measurements may be performed rapidly and on site, which is important for effective quality control and a prerequisite for processes monitoring. An important advantage as compared to traditional methods is the possibility of simultaneous determination of different chemical components or physical properties in the sample from a single spectral measurement.

Acknowledgements

Grant NN312428239, 2010–2013, from the Polish Ministry of Science and Higher Education is gratefully acknowledged.

References

Ahmed, M.K., Daun, J.K. and Przybylski, R. 2005. FT-IR based methodology for quantitation of total tocopherols, tocotrienols and plastochromanol-8 in vegetable oils. Journal of Food Composition and Analysis. 18: 359–364.

Armenta, S., Moros, J., Garrigues, S. and Guardia, M.D.L. 2010a. Chapter 58—Determination of Olive Oil Parameters by Near Infrared Spectrometry. In: Preedy, V.R. and Watson, R.R. (eds.) Olives and Olive Oil in Health and Disease Prevention. San Diego: Academic Press.

Armenta, S., Moros, J., Garrigues, S. and Guardia, M.D.L. 2010b. The use of near-infrared spectrometry in the olive oil industry. Critical Reviews in Food Science and Nutrition. 50: 567–582.

Azizian, H. and Kramer, J.G. 2005. A rapid method for the quantification of fatty acids in fats and oils with emphasis on trans fatty acids using fourier transform near infrared spectroscopy (FT-NIR). Lipids. 40: 855–867.

Baeten, V., Fernandez Pierna, J.A., Dardenne, P., Meurens, M., Garcia-Gonzalez, D.L. and Aparicio-Ruiz, R. 2005. Detection of the presence of hazelnut oil in olive oil by FT-Raman and FT-MIR Spectroscopy. Journal of Agricultural and Food Chemistry. 53: 6201–6206.

Bakeev, K.A. 2005. Process Analytical Technology, Blackwell Publishing Ltd.

Belitz, H.-D., Grosch, W. and Schieberle, P. 2009. Food Chemistry, Springer.

Berrueta, L.A., Alonso-Salces, R.M. and Héberger, K. 2007. Supervised pattern recognition in food analysis. Journal of Chromatography A. 1158: 196–214.

Bevilacqua, M., Bucci, R., Magri, A.D., Magri, A.L. and Marini, F. 2012. Tracing the origin of extra virgin olive oils by infrared spectroscopy and chemometrics: a case study. Anal. Chim. Acta. 717: 39–51.

Bro, R. 2003. Multivariate calibration: What is in chemometrics for the analytical chemist? Analytica Chimica Acta. 500: 185–194.
Casale, M., Oliveri, P., Casolino, C., Sinelli, N., Zunin, P., Armanino, C., Forina, M. And Lanteri, S. 2012. Characterisation of PDO olive oil Chianti classico by non-selective (UV-visible, NIR and MIR spectroscopy) and selective (fatty acid composition) analytical techniques. Anal. Chim. Acta. 712: 56–63.
Cerretani, L., Giuliani, A., Maggio, R.M., Bendini, A., Toschi, T.G. and Cichelli, A. 2010. Rapid FTIR determination of water, phenolics and antioxidant activity of olive oil. European Journal of Lipid Science and Technology. 112: 1150–1157.
Commission Regulation (EU) No 61/2011, L23 amending Regulation (EEC) No 2568/91 on the characteristics of olive oil and olive-residue oil and on the relevant methods of analysis. Bruxelles, Belgium: Official Journal of European Union, Publications Office of the European Union, The European Commission.
Cozzolino, D., Cynkar, W.U., Shah, N. and Smith, P. 2011. Multivariate data analysis applied to spectroscopy: Potential application to juice and fruit quality. Food Research International. 44: 1888–1894.
Dankowska, A. and Malecka, M. 2009. Application of synchronous fluorescence spectroscopy for determination of extra virgin olive oil adulteration. European Journal of Lipid Science and Technology. 111: 1233–1239.
De la Mata, P., Dominguez-Vidal, A., Bosque-Sendra, J.M., Ruiz-Medina, A., Cuadros-Rodríguez, L. and Ayora-Cañada, M.J. 2012. Olive oil assessment in edible oil blends by means of ATR-FTIR and chemometrics. Food Control. 23: 449–455.
Diaz, T.G., Meras, I.D., Correa, C.A., Roldan, B. and Caceres, M.I.R. 2003. Simultaneous fluorometric determination of chlorophylls a and b and pheophytins a and b in olive oil by partial least-squares calibration. Journal of Agricultural and Food Chemistry. 51: 6934–6940.
El-Abassy, R.M., Donfack, P. and Materny, A. 2009. Visible Raman spectroscopy for the discrimination of olive oils from different vegetable oils and the detection of adulteration. Journal of Raman Spectroscopy. 40: 1284–1289.
El-Abassy, R.M., Donfack, P. and Materny, A. 2010. Assessment of conventional and microwave heating induced degradation of carotenoids in olive oil by VIS Raman spectroscopy and classical methods. Food Research International. 43: 694–700.
Ellis, D.I., Brewster, V.L., Dunn, W.B., Allwood, J.W., Golovanov, A.P. and Goodacre, R. 2012. Fingerprinting food: current technologies for the detection of food adulteration and contamination. Chemical Society Reviews. 41: 5706–5727.
Engel, J., Gerretzen, J., Szymańska, E., Jansen, J.J., Downey, G., Blanchet, L. and Buydens, L.M.C. 2013. Breaking with trends in pre-processing? TrAC Trends in Analytical Chemistry. 50: 96–106.
Galtier, O., Dupuy, N., Le Dreau, Y., Ollivier, D., Pinatel, C., Kister, J. and Artaud, J. 2007. Geographic origins and compositions of virgin olive oils determinated by chemometric analysis of NIR spectra. Anal. Chim. Acta. 595: 136–44.
Gómez-Caravaca, A.M., Maggio, R.M., Verardo, V., Cichelli, A. and Cerretani, L. 2013. Fourier transform infrared spectroscopy–Partial Least Squares (FTIR–PLS) coupled procedure application for the evaluation of fly attack on olive oil quality. LWT-Food Science and Technology. 50: 153–159.
Guilbault, G.G. 1991. Practical Fluorescence, New York, Marcel Dekker, Inc.
Guillén, M.D. and Cabo, N. 1997. Infrared spectroscopy in the study of edible oils and fats. Journal of the Science of Food and Agriculture. 75: 1–11.
Guimet, F., Boqué, R. and Ferré, J. 2004. Cluster analysis applied to the exploratory analysis of commercial Spanish olive oils by means of excitation-emission fluorescence spectroscopy. Journal of Agricultural and Food Chemistry. 52: 6673–6679.
Guimet, F., Ferré, J. and Boqué, R. 2005a. Rapid detection of olive-pomace oil adulteration in extra virgin olive oils from the protected denomination of origin "Siurana" using

excitation-emission fluorescence spectroscopy and three-way methods of analysis. Analytica Chimica Acta. 544: 143–152.

Guimet, F., Ferre, J., Boque, R., Vidal, M. and Garcia, J. 2005b. Excitation-emission fluorescence spectroscopy combined with three-way methods of analysis as a complementary technique for olive oil characterization. Journal of Agricultural and Food Chemistry. 53: 9319–9328.

Gurdeniz, G. and Ozen, B. 2009. Detection of adulteration of extra-virgin olive oil by chemometric analysis of mid-infrared spectral data. Food Chemistry. 116: 519–525.

Hibbert, D.B., Minkkinen, P., Faber, N.M. and Wise, B.M. 2009. IUPAC project: A glossary of concepts and terms in chemometrics. Analytica Chimica Acta. 642: 3–5.

Inarejos-García, A.M., Gómez-Alonso, S., Fregapane, G. and Salvador, M.D. 2013. Evaluation of minor components, sensory characteristics and quality of virgin olive oil by near infrared (NIR) spectroscopy. Food Research International. 50: 250–258.

Javidnia, K., Parish, M., Karimi, S. and Hemmateenejad, B. 2013. Discrimination of edible oils and fats by combination of multivariate pattern recognition and FT-IR spectroscopy: A comparative study between different modeling methods. Spectrochimica Acta Part A: Molecular and Biomolecular Spectroscopy. 104: 175–181.

Kjeldahl, K. and Bro, R. 2010. Some common misunderstandings in chemometrics. Journal of Chemometrics. 24: 558–564.

Korifi, R., Le Dréau, Y., Molinet, J., Artaud, J. and Dupuy, N. 2011. Composition and authentication of virgin olive oil from French PDO regions by chemometric treatment of Raman spectra. Journal of Raman Spectroscopy. 42: 1540–1547.

Kuligowski, J., Carrión, D., Quintás, G., Garrigues, S. and De la Guardia, M. 2012. Direct determination of polymerised triacylglycerides in deep-frying vegetable oil by near infrared spectroscopy using Partial Least Squares regression. Food Chemistry. 131: 353–359.

Lakowicz, J.R. 2006. Principles of Fluorescence Spectroscopy, New York, Springer-Verlag New York Inc.

Lerma-García, M.J., Ramis-Ramos, G., Herrero-Martínez, J.M. and Simó-Alfonso, E.F. 2010. Authentication of extra virgin olive oils by Fourier-transform infrared spectroscopy. Food Chemistry. 118: 78–83.

Lerma-García, M.J., Simó-Alfonso, E.F., Bendini, A. and Cerretani, L. 2011. Rapid evaluation of oxidised fatty acid concentration in virgin olive oil using Fourier-transform infrared spectroscopy and multiple linear regression. Food Chemistry. 124: 679–684.

Lloyd, J.B.F. 1971. Synchronized excitation of fluorescence emission spectra. Nature-Physical Science. 231: 6–65.

Long, D.A. 2002. The Raman Effect: A Unified Treatment of the Theory of Raman Scattering by Molecules, New York, John Wiley & Sons, Ltd.

López-Díez, E.C., Bianchi, G. and Goodacre, R. 2003. Rapid quantitative assessment of the adulteration of virgin olive oils with hazelnut oils using raman spectroscopy and chemometrics. Journal of Agricultural and Food Chemistry, 51: 6145–6150.

Luna, A.S., Da Silva, A.P., Ferre, J. and Boque, R. 2013a. Classification of edible oils and modeling of their physico-chemical properties by chemometric methods using mid-IR spectroscopy. Spectrochim Acta A Mol. Biomol. Spectrosc. 100: 109–14.

Luna, A.S., Da Silva, A.P., Pinho, J.S., Ferre, J. and Boque, R. 2013b. Rapid characterization of transgenic and non-transgenic soybean oils by chemometric methods using NIR spectroscopy. Spectrochim Acta A Mol. Biomol. Spectrosc. 100: 115–9.

Maggio, R.M., Kaufman, T.S., Carlo, M.D., Cerretani, L., Bendini, A., Cichelli, A. and Compagnone, D. 2009. Monitoring of fatty acid composition in virgin olive oil by Fourier transformed infrared spectroscopy coupled with partial least squares. Food Chemistry. 114: 1549–1554.

Mailer, R. 2004. Rapid evaluation of olive oil quality by NIR reflectance spectroscopy. Journal of the American Oil Chemists' Society. 81: 823–827.

Marquez, A.J., Díaz, A.M. and Reguera, M.I.P. 2005. Using optical NIR sensor for on-line virgin olive oils characterization. Sensors and Actuators B: Chemical. 107: 64–68.
McClure, W.F. 2003. 204 Years of Near Infrared Technology, 1800-2004. Journal of Infrared Spectroscopy. 11: 487–518.
Moros, J., Garrigues, S. and Guardia, M.D.L. 2010. Vibrational spectroscopy provides a green tool for multi-component analysis. TrAC Trends in Analytical Chemistry. 29: 578–591.
Moyano, M.J., Meléndez-Martínez, A.J., Alba, J. and Heredia, F.J. 2008. A comprehensive study on the colour of virgin olive oils and its relationship with their chlorophylls and carotenoids indexes (I): CIEXYZ non-uniform colour space. Food Research International. 41: 505–512.
Ndou, T.T. and Warner, I.M. 1991. Applications of multidimensional absorption and luminescence spectroscopies in analytical chemistry. Chemical Reviews. 91: 493–507.
Ng, C.L., Wehling, R.L. and Cuppett, S.L. 2006. Method for determining frying oil degradation by near-infrared spectroscopy. Journal of Agricultural and Food Chemistry. 55: 593–597.
Nicolaï, B.M., Beullens, K., Bobelyn, E., Peirs, A., Saeys, W., Theron, K.I. and Lammertyn, J. 2007. Nondestructive measurement of fruit and vegetable quality by means of NIR spectroscopy: A review. Postharvest Biology and Technology. 46: 99–118.
Nielsen, S.S. 2010. Food Analysis/edited by S. Suzanne Nielsen, New York ; London, Springer.
Oliveri, P. and Downey, G. 2012. Multivariate class modeling for the verification of food-authenticity claims. TrAC Trends in Analytical Chemistry. 35: 74–86.
Otto, M. 2007. Chemometrics, WILEY-VCH Verlag GmbH.
Pereira, A.F.C., Pontes, M.J.C., Neto, F.F.G., Santos, S.R.B., Galvão, R.K.H. and Araújo, M.C.U. 2008. NIR spectrometric determination of quality parameters in vegetable oils using iPLS and variable selection. Food Research International. 41: 341–348.
Poulli, K.I., Chantzos, N.V., Mousdis, G.A. and Georgiou, C.A. 2009. Synchronous fluorescence spectroscopy: tool for monitoring thermally stressed edible oils. Journal of Agricultural and Food Chemistry. 57: 8194–8201.
Sherazi, S.T.H., Arain, S., Mahesar, S.A., Bhanger, M.I. and Khaskheli, A.R. 2013. Erucic acid evaluation in rapeseed and canola oil by Fourier transform-infrared spectroscopy. European Journal of Lipid Science and Technology. 115: 535–540.
Sikorska, E., Gliszczynska-Swiglo, A., Khmelinskii, I. and Sikorski, M. 2005. Synchronous fluorescence spectroscopy of edible vegetable oils. Quantification of tocopherols. Journal of Agricultural and Food Chemistry. 53: 6988–6994.
Sikorska, E., Khmelinskii, I. and Sikorski, M. 2012. Analysis of olive oils by fluorescence spectroscopy: Methods and applications. *In*: Boskou, D. (ed.). Olive Oil—Constituents, Quality, Health Properties and Bioconversions. InTech.
Sinelli, N., Casale, M., Di Egidio, V., Oliveri, P., Bassi, D., Tura, D. and Casiraghi, E. 2010a. Varietal discrimination of extra virgin olive oils by near and mid infrared spectroscopy. Food Research International. 43: 2126–2131.
Sinelli, N., Cerretani, L., Egidio, V.D., Bendini, A. and Casiraghi, E. 2010b. Application of near (NIR) infrared and mid (MIR) infrared spectroscopy as a rapid tool to classify extra virgin olive oil on the basis of fruity attribute intensity. Food Research International. 43: 369–375.
Small, G.W. 2006. Chemometrics and near-infrared spectroscopy: Avoiding the pitfalls. TrAC Trends in Analytical Chemistry. 25: 1057–1066.
Sun, D.W. 2009. Infrared Spectroscopy for Food Quality Analysis and Control, Academic Press.
Tapp, H.S., Defernez, M. and Kemsley, E.K. 2003. FTIR Spectroscopy and multivariate analysis can distinguish the geographic origin of extra virgin olive oils. Journal of Agricultural and Food Chemistry. 51: 6110–6115.

Torrecilla, J.S., Rojo, E., Domínguez, J.C. and Rodríguez, F. 2010. A novel method to quantify the adulteration of extra virgin olive oil with low-grade olive oils by UV–Vis. Journal of Agricultural and Food Chemistry. 58: 1679–1684.

Valli, E., Bendini, A., Maggio, R.M., Cerretani, L., Toschi, T.G., Casiraghi, E. and Lercker, G. 2013. Detection of low-quality extra virgin olive oils by fatty acid alkyl esters evaluation: a preliminary and fast mid-infrared spectroscopy discrimination by a chemometric approach. International Journal of Food Science & Technology. 48: 548–555.

Van de Voort, F.R., Sedman, J. and Russin, T. 2001. Lipid analysis by vibrational spectroscopy. European Journal of Lipid Science and Technology. 103: 815–826.

Vlachos, N., Skopelitis, Y., Psaroudaki, M., Konstantinidou, V., Chatzilazarou, A. and Tegou, E. 2006. Applications of Fourier transform-infrared spectroscopy to edible oils. Anal. Chim. Acta. 573–574: 459–65.

Wold, S., Sjostrom, M. and Eriksson, L. 2001. PLS-regression: a basic tool of chemometrics. Chemometrics and Intelligent Laboratory Systems. 58: 109–130.

Woodcock, T., Downey, G. and O'Donnell, C.P. 2008. Confirmation of declared provenance of European extra virgin olive oil samples by NIR spectroscopy. Journal of Agricultural and Food Chemistry. 56: 11520–11525.

Xiaobo, Z., Jiewen, Z., Povey, M.J.W., Holmes, M. and Hanpin, M. 2010. Variables selection methods in near-infrared spectroscopy. Analytica Chimica Acta. 667: 14–32.

Yang, H., Irudayaraj, J. and Paradkar, M. 2005. Discriminant analysis of edible oils and fats by FTIR, FT-NIR and FT-Raman spectroscopy. Food Chemistry. 93: 25–32.

Zandomeneghi, M., Carbonaro, L. and Caffarata, C. 2005. Fluorescence of vegetable oils: Olive oils. Journal of Agricultural and Food Chemistry. 53: 759–766.

Zhang, X.-F., Zou, M.-Q., Qi, X.-H., Liu, F., Zhang, C. and Yin, F. 2011. Quantitative detection of adulterated olive oil by Raman spectroscopy and chemometrics. Journal of Raman Spectroscopy. 42: 1784–1788.

Index

A

absorption 201–205, 207, 208, 210, 211, 229
alkaloids 110, 111, 113
anthocyanins 113, 114, 116–118, 120–122, 124–128

B

back extrusion 78, 79, 91–93
bending 205, 206, 209
betalains 113, 114, 118, 120
blade cell 83
Bligh and Dyer 63, 64, 153–157, 162, 163, 191

C

calibration 201, 202, 214–216, 218, 220–223, 230
capillary electrophoresis 72, 125
carotenoids 110–112, 115, 119–123, 125, 127–130
chemometrics 201–234
chewiness 21, 76, 77, 96, 98–100, 104
chlorophylls 112, 115, 116, 119, 121, 123, 124, 128–130
chroma 55
chromatographic methods 119
chromophores 201
CIELab 55
cluster analysis 214
cohesiveness 19–21, 88, 92, 93, 96–100
color 44–61, 110–114, 125, 127
colorimeter 56, 57
compression 3, 7–10, 12–15, 17–20, 26, 76, 78–80, 85, 86, 91, 93–97, 99–103
computer analysis 44, 45, 56
counter current chromatography 124

D

deformation 1–4, 10, 11, 14–16, 20, 28, 205, 206, 208

F

fatty acid 64, 66, 70, 71
finger method 78, 103
fish 76–109, 146, 155, 156, 158, 160, 168, 169, 179, 186, 189
fluorescence 210, 211, 215, 226, 228
Folch method 153, 155, 161, 163, 191
fracturability 77, 88, 96–99, 104
fruit 33, 62–75, 110–141

G

gas chromatography 65–67, 71
gravimetric 155–160

H

Hara and Radin method 162
hardness 7, 17, 19–21, 77, 79, 86–88, 91, 96–101
hemoglobin 45
high performance liquid chromatography 68, 69, 119
hue angle 55
Hunter Lab 55, 56

I

illuminant 55, 56
infrared 201–205, 207, 229
invasive methods 126, 127

J

jam 6, 23, 28, 34, 35
Jensen method 154, 157, 163, 164
juice 6, 33, 34

K

Kramer cell 78, 88, 93–95

L

lightness 55, 56
lipid 62–65, 68–72, 142–200

M

mass spectrometry 65, 67–70
meat 44–57, 76–109, 142–200
metmyoglobin 50, 51
microwave-assisted extraction 160, 167
muscle 44–61
myoglobin 44–46, 48–50

N

near infrared spectroscopy 154, 159, 169
non-chromatographic techniques 126
non-organic solvents 166
nuclear magnetic resonance 71

O

observer angle 55
oil 63, 64, 66–72, 201, 202, 204, 205, 207, 209–212, 223–230
omega 3 142–144, 146–148, 150, 151, 186
open column chromatography 122
Ottawa cell 78, 93, 96
overtone 204, 206–208
oxymyoglobin 50

P

paste 28, 37, 38
penetration 19
phospholipids 62, 63, 65, 68, 70
pigments 44–61, 110–141
polyphenolic compounds 110, 111
polyunsaturated fatty acids 142, 168, 180, 183
porphyrin 49
principal component analysis 213, 214
pulp 6, 34–39
puree 35–37, 39

R

radiation 202–204, 210, 217, 229
Raman 201–204, 208, 209, 225–229
Raman spectroscopy 71
razor blade 78, 87–89

reflectance 44, 45, 52, 53, 55–57
resilience 79, 96, 98, 100
rheology 1, 2, 4, 5, 22–24, 26, 32, 33, 39

S

saturated fatty acids 142, 144, 145, 150, 168, 177, 180, 183
seaweeds 142, 148, 149
shear rate 1, 4–6, 22–25, 27, 31, 32, 35–39
shear strain 4, 7, 21
shear stress 1, 3–7, 13, 21–25, 29, 31, 32, 36–39
shearing test 78, 80, 94
silica 64, 68
Smedes method 156, 163
Soxhlet 63, 153, 157, 160, 165–167
spectra 201–218, 223, 229
springiness 96, 98–100
stretching 204–209
supercritical fluid chromatography 124
supercritical fluid extraction 154, 158

T

tension 3, 7, 15–17, 19, 30
tetrapyrroles 110–112
texture 1–3, 7, 13, 17, 19, 20, 34, 39
texture profile analysis 19, 20, 85, 86, 96, 98–101
thin-layer chromatography 64, 65, 122
triacylglycerols 68, 152, 168, 175, 184, 188, 207, 224
triglycerides 62–64, 71

U

ultraviolet 201–204, 210
unsaturated fatty acids 142, 144, 150, 168, 177, 180, 183

V

vegetable 1–43, 62–75, 110–141
vibrational spectroscopy 203, 223, 229
viscosity 76, 77, 91–93, 100, 104
visible 201–204, 210, 211
visual evaluation 44, 45, 54
Volodkevich bite jaws 78, 90

W

Warner-Bratzler 78, 80, 81
Weibull-Stoldt method 160
Werner-Schmid method 159

Chapter 2

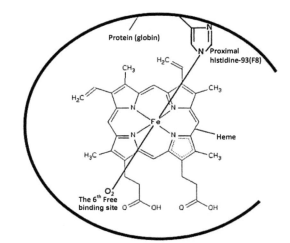

Figure 2.1 Myoglobin structure.

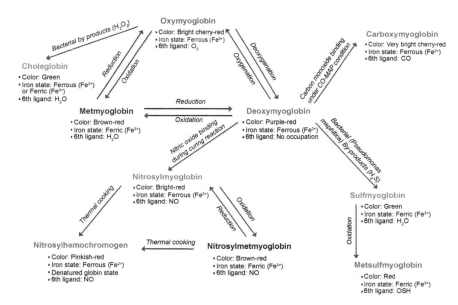

Figure 2.2 Visible myoglobin redox interconversions under various oxidative, reducing, microbiologicalconditions and different chemical states of myoglobin.

Chapter 4

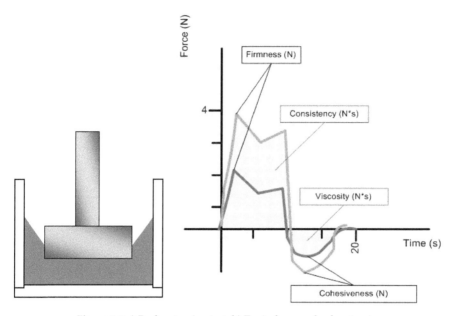

Figure 4.6 a) Back extrusion test; b) Typical curves back extrusion.

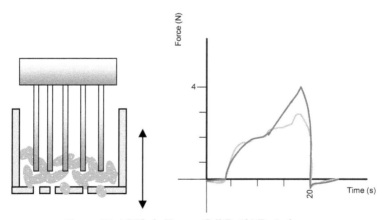

Figure 4.7 a) 5-Blade Kramer Cell Test b) Typical curve.

240 *Methods in Food Analysis*

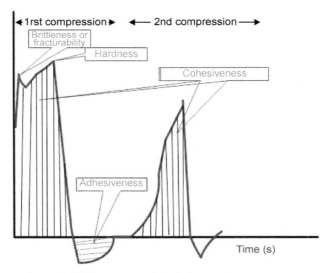

Figure 4.10 TPA curves and typical parameters.

Ingram Content Group UK Ltd.
Milton Keynes UK
UKHW022157030523
421200UK00006B/32